Warwick A. Smith

C Programming
for Embedded Microcontrollers

C Programming
for Embedded Microcontrollers

Warwick A. Smith

Elektor International Media BV
PO Box 11
6114 ZG Susteren
The Netherlands

British Library Cataloguing in Publication Data
A catalogue record for this book is available from the British Library

ISBN 978-0-905705-80-4

Prepress production: Autronic, Blaricum
Cover design: Helfrich Ontwerpbureau, Deventer First published in the United Kingdom 2008, Fourth Edition 2013
Printed in the Netherlands by Wilco, Amersfoort

089029-1/UK

Table of Contents

Introduction

This book will teach you the C programming language. It will also teach you how to use the C programming language to program embedded systems.

Learning a new programming language and learning how to program embedded systems covers a lot of ground as there are many things to learn: the C programming language itself, some knowledge of microcontroller architecture, some electronics, microcontroller peripherals, number systems and logic. It may seem to be a daunting task to learn about all these aspects of an embedded system and programming language, but taking it one step at a time as this book does makes it easy, enjoyable and a lot of fun.

This book concentrates on ARM microcontrollers and only mentions other microcontrollers in passing. The C language applies equally to other microcontrollers, but when learning to program, a particular architecture must be chosen as hardware differences between microcontrollers will change how they are programmed.

The ARM Architecture

ARM stands for **A**dvanced **R**ISC **M**achine and is a 32-bit processor architecture. There are a number of microcontroller manufacturers that use the ARM architecture in their microcontrollers. These microcontrollers contain an "ARM core" that is interfaced to memory and peripheral devices and packaged in a single chip.

RISC stands for **R**educed **I**nstruction **S**et **C**omputer and is a processor that is designed with a simplified instruction set in order to produce higher performance. The ARM architecture has been designed this way.

Target Audience

This book has been written for the hobbyist, student and engineer wanting to learn the C programming language in an embedded environment using microcontrollers.

What is an Embedded System?

An embedded system is an electronic circuit board that contains a microprocessor (µP) or microcontroller (µC) and other electronic parts, running a software program and used for a specific purpose rather than a general purpose as a Personal Computer (PC) does. (µ = micro.)

An example of an embedded system is a calculator. It is specifically made to perform mathematical calculations input by a user on the keypad – it is a dedicated device. A PC on the other hand is a general purpose computer that can be used for any number of purposes such as word processing, playing games, a file server, etc. (i.e. not an embedded system)

Other examples of embedded systems are cell phones and computer keyboards as they contain embedded microcontrollers that give these devices their functionality. A washing machine may contain an embedded system that responds to button presses on its front panel and controls washing cycles.

A **microprocessor** is a silicon chip (also known as an Integrated Circuit or I.C.) containing a Central Processing Unit (CPU) that will fetch program instructions from memory and execute (run) them. The memory required by the microprocessor is external to the microprocessor chip – it must be put on the circuit board by the hardware engineer designing the system, whether it is to be used as an embedded system or general purpose computer. There are a number of microprocessor architectures such as Intel's x86 architecture, Motorola's (Freescale's) Power PC architecture or the ARM architecture.

A **microcontroller** contains a microprocessor and a number of peripheral devices such as timers, serial ports, general purpose input/output (I/O) pins, counters, analogue inputs, etc. all inside a single silicon chip. A microcontroller will usually have memory on the chip as well. Some microcontrollers will be able to support external memory and others have enough memory on-chip and do not have the ability to support external memory. Microcontrollers are made to be used in embedded systems as their on-chip peripherals and memory enable an embedded system designer to save circuit board space by not having to add these things as external devices on the circuit board.

Two ARM Microcontrollers, one from NXP (Philips) (left) and one from Atmel (right)

The hobbyist, student or engineer will typically choose a microcontroller that he would like to use and then build or buy his embedded system containing the chosen microcontroller. A number of evaluation boards (EVB) are available for this purpose. An evaluation board is a Printed Circuit Board (PCB) containing a microcontroller that is connected to external connectors and Integrated Circuits (I.C.s) for interfacing to RS-232 serial ports, USB ports, Light Emitting Diodes (LEDs), switches, etc. so that the microcontroller can be programmed and evaluated without the user having to first build the hardware.

This book takes the approach of using a pre-made evaluation board or board made for hobbyists as an embedded system. You will learn to write programs that run on an embedded system and switch LEDs on and off, read switches to see if they are being pressed or released, send messages out of the serial port to be displayed on a PC and more.

Why use the C Programming Language?

If I think of all the technology changes that have taken place in the world of microelectronics since the start of my career as an embedded programmer, the one thing that has stayed the same is the C language. Processor architectures and systems have come and gone, but the common thing between all of them is that they were programmed in C.

The reason for the popularity and use of C is that C is a standard language and ideally suited to embedded programming where access to the hardware of a system is required. C programming tools produce small and fast programs – a requirement when you have limited amounts of memory available as on an embedded system.

Whenever a new processor architecture comes on the market, you can be pretty sure that there will be C programming tools available for it.

You may have heard that C is cryptic and difficult to learn, but this is nonsense. C is a simple language with very few keywords. It is easy to learn. As with any new language, you need to become familiar with the syntax of the language. This book will explain everything to you, starting with simple and easy programs. When I leaned to program in C, I thoroughly enjoyed it and I hope that you will too.

Prerequisites

In order to use this book you will need to be able to operate a computer running Windows or know someone who is able to show you how.

Any knowledge of electronics will help. For example you should know what an LED is, a switch, I.C. and a resistor. I will refer to basic electronic components such as these in the text. If you don't know what these items are, there are plenty of books available for teaching you basic electronics.

Any knowledge of programming in any language will help, but is not required.

Basic mathematics is also assumed. i.e. addition, subtraction, multiplication, division, counting and a little algebra. If you are not good at mathematics, that is not a problem as the mathematics is very basic and the programs that you will write will actually help you to understand mathematics better.

You will need to have an Internet connection in order to download software programs and example programs used with this book (all software used is free). It is assumed that you know how to download and install software on a Windows PC.

With the above requirements met you will be able to work through chapters 1 to 7. After chapter 7 you will need to have an embedded system available for programming.

Here is what you will need in addition to a PC and Internet connection:

Shopping List:

Quantity	Device	Notes
1	Embedded system	Details given below and in chapter 7.
1	RS-232 port on PC	Already on PC or if not, buy a USB to RS-232 adapter. Details are in chapter 10.
1	RS-232 cable	Connects embedded system to PC serial port. Details in chapter 10.
1	Standard USB cable	If you choose the recommended embedded system. Other embedded systems may need it too.
1	JTAG ICE	Optional hardware tool used for debugging. Details in chapter 15.

A summary of prerequisites and more details are given below.

List of prerequisites (hardware and software):

For chapters 1 to 7

1. A PC running Windows 2000 or later.

2. An Internet connection – preferably broadband.

For chapters 8 and onwards – in addition to the above

1. An embedded system – preferably containing an ARM7 microcontroller as described below.

2. Serial RS-232 port on the PC or USB to RS-232 serial port adapter.

3. RS-232 serial cable to link the embedded system to the PC (crossed-over (null modem) or straight cable will depend on how the embedded system is wired).

4. A standard USB cable (may be included with your embedded system when bought).

5. An optional JTAG ICE for debugging.

Embedded system requirements:

Although it is possible to use a number of different embedded systems with this book the following are required to be on the embedded system circuit board:

1. The embedded system must have LEDs that the user can control with a program (four or more).

2. It must have switches that can be read by a program.

3. It must have an RS-232 serial port.

4. It must have an analogue to digital converter (ADC).

It is preferable to have an ARM7 microcontroller on the embedded system board and specifically the AT91SAM7S256 from Atmel or a similar microcontroller from the same family (The AT91SAM7S microcontroller family members differ from each other in the amount of memory that they have on-chip).

Choosing this specific microcontroller is because microcontrollers are different from each other in terms of the number of built in peripherals that they contain, the amount of memory that they contain and the programming tools used to program them. You can choose a different microcontroller, but I highly recommend getting the above one. The following points will help to explain why:

1. The free open source tools used in this book work with any ARM7 microcontroller so you could get one from Atmel or NXP or any other manufacturer that uses the ARM7 processor core in their microcontrollers. If you choose a different microcontroller architecture, you will need a different set of software tools for programming it, so you will have to find out how to set them up and use them yourself. You may experience some difficulty here, depending on the quality of documentation available for the microcontroller/evaluation board that you choose. You will be on your own here because this book can't possibly cover every available microcontroller out there.

2. Even if you choose a different ARM7 microcontroller from the one used in this book, different manufacturers have different requirements for setting up and using the on-chip peripheral devices. So to program a serial port on the Atmel ARM7 requires different programming instructions than those used to program a serial port on an NXP ARM7 microcontroller for example (although the microcontroller core (ARM7) and programming tools are the same).

3. There may not be free or open source programming tools available for a different microcontroller architecture.

4. As a result of the above points, the chapters covering setup of the programming tools will be useless to you unless you are using the specified microcontroller architecture (ARM7).

5. Microcontroller programming requires that you program hardware directly and as each microcontroller has different hardware, you would have to read what this book says about the chosen microcontroller and try to convert it to use on your different microcontroller – expect to do a lot of panel-beating of the code, so choosing a different microcontroller is only recommended for advanced users.

On the other hand using the recommended Atmel AT91SAM7S256 microcontroller has the following advantages:

1. It uses a 32-bit ARM7 core = fast and price comparable to 8-bit microcontrollers with similar peripheral hardware.

2. Free open source GNU programming tools are available for programming it.

3. Commercial programming tools are available from a number of vendors if required.

4. The AT91SAM7S256 has an on-chip boot loader and PC program that work together to enable you to load programs into the chip without the need for a special programmer – just use a standard USB cable. Make sure that the embedded system that you buy that uses this microcontroller has its hardware set up to use this feature.

5. There is more than enough memory on the microcontroller to run all of the examples in this book. Tiny 8-bit microcontrollers with minimal memory will have a problem here.

6. Atmel has made an effort to support the open source GNU tools and there are application notes and tutorials available from their website.

7. This microcontroller is reasonably priced and getting an evaluation board is affordable.

8. The microcontroller is very popular and there are lots of projects available on the Internet that use it.

9. A number of evaluation boards are available from different manufacturers.

10. Some evaluation boards are powered by the USB connection and so no power supply needs to be bought.

11. Single step debugging is possible with affordable JTAG ICE adapters.

Chapter 7 in this book will explain what to look for when choosing your embedded system.

Debugging

Debugging refers to finding and fixing program errors. A programmer will write a program and then test it to see if it works as intended. If there is a problem with the program, e.g. "Why does the LED not switch on when it is supposed to?", the programmer will debug the program.

Many microcontrollers have a JTAG port that is wired to a connector on the embedded system board. A hardware device called a JTAG ICE is plugged into the JTAG port, connecting the embedded system to a PC. When a software program called a debugger is run on the PC, it accesses the embedded system through the JTAGE ICE and enables the programmer to step through the program and examine the contents of the microcontroller's memory. This helps the programmer to find the bug (program error) and fix it.

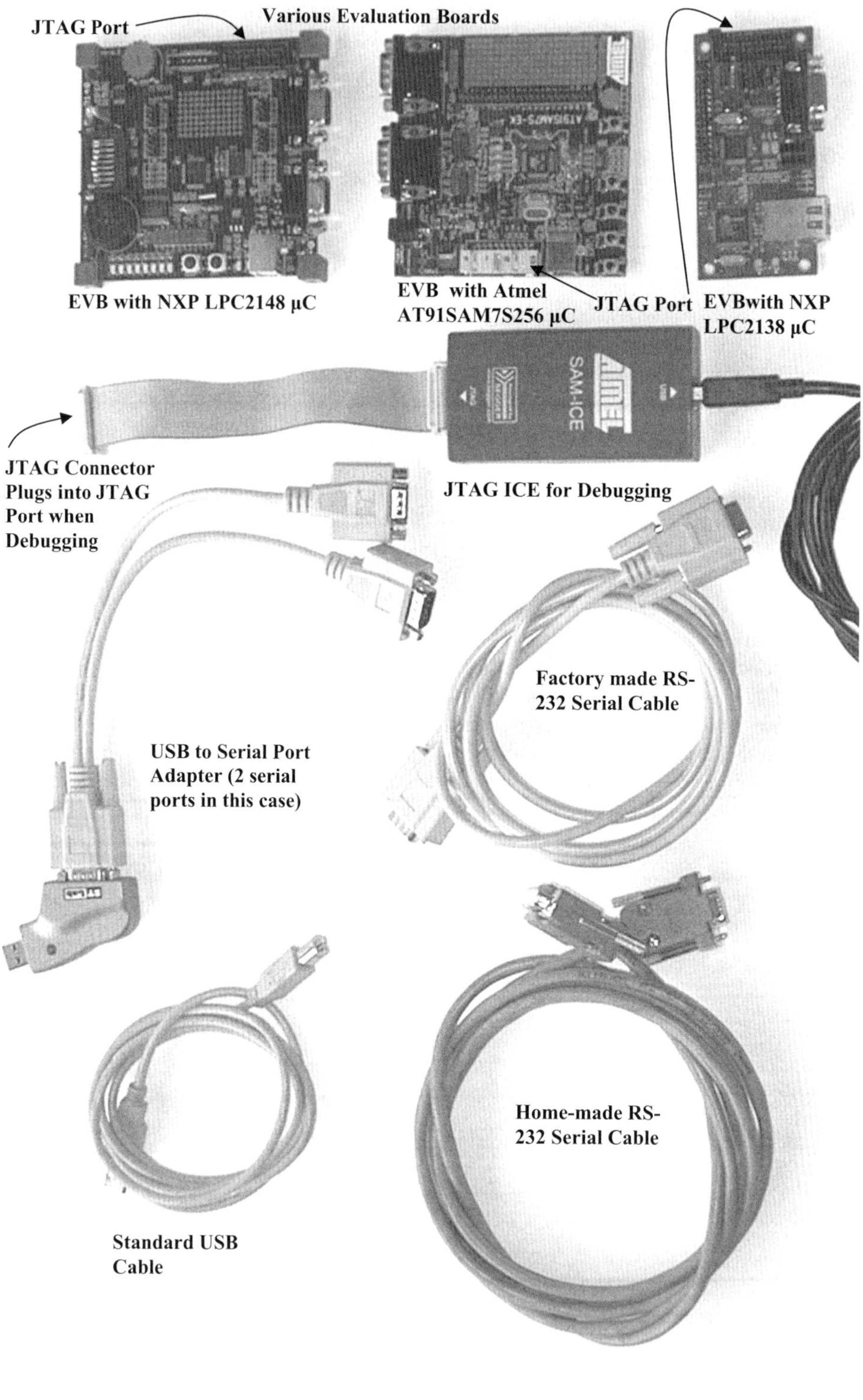
JTAG Port
Various Evaluation Boards
EVB with NXP LPC2148 µC
EVB with Atmel AT91SAM7S256 µC
JTAG Port
EVBwith NXP LPC2138 µC
SAM-ICE
JTAG Connector Plugs into JTAG Port when Debugging
JTAG ICE for Debugging
Factory made RS-232 Serial Cable
USB to Serial Port Adapter (2 serial ports in this case)
Home-made RS-232 Serial Cable
Standard USB Cable

The Approach Taken in this Book

This book uses free or open source software only. You will start learning to program right from chapter 1 after installing and setting up programming tools.

The first 7 chapters start teaching you to program in C on a PC while also increasing your knowledge of embedded systems. Using the PC enables you to see how programming works without the hassle of having to understand the hardware of the system. You will also learn about embedded system hardware, memory, number systems and logic operations that are all necessary in order to understand how an embedded system is programmed.

Once you have mastered some basic C programming skills, chapter 8 will introduce you to embedded programming. You will move your programming skills over to an embedded system and then continue to learn the C programming language on the embedded system. How to program various hardware devices on the embedded system will then be covered.

Some Good Advice

Don't rush through the first 7 chapters to get to the embedded programming. Take your time to learn and understand how the PC programs work before moving between chapters. I know that you bought this book in order to learn to program embedded systems and not PCs but learning parts of the C language on the PC will help you to concentrate on how the C language works without having to know how the hardware works.

You only need to buy your embedded system after chapter 7, so take some time to choose it. You can even read past chapter 7 so that you will understand what is involved in programming an embedded system.

If you can't type, learn to type while learning C. This is what I did when learning to program and have never regretted it. I always entered the example programs by typing them in rather than just running the example programs from disk (or download).

Learning C and embedded systems is great fun and a big adventure, so it is my wish that you will thoroughly enjoy using this book and learning new things.

1 Your First C Program

What you will do in this chapter:

- Download and install some programming tools
- Set up your PC to run the programming tools
- Install a text editor
- Write your first C program
- Compile and run your first program

1.1 About the Programming Tools

In order to start programming in the C programming language, you will need to install some programming tools (software) onto your computer. You will start by installing tools that allow you to create programs that will run on the PC itself. It is much simpler to program the PC than to start programming an embedded system, so I would like to introduce the C programming language to you by using the PC. Later you will move your programming skills over to embedded systems, as already mentioned in the introduction.

The tools are usually referred to as the "C compiler" although they contain several tools besides the compiler. Most Linux distributions come with the GCC C compiler on the installation CD and will most likely install the tools by default. We will be dealing with installing the programming tools on a Windows PC.

We will install the GCC C compiler for Windows called DJGPP. This compiler is a "command line compiler" in that it does not have a fancy Graphical User Interface (GUI), but runs from the command prompt in a "DOS box" or "Windows console". The first programs that we write will run in "console" mode in order to isolate you from Windows GUI programming. In console mode you can write much simpler C programs without having to worry about windows, buttons and other graphical objects. The compiler will produce 32-bit console mode programs.

The next few sections in this chapter will show you how to download, install and setup the DJGPP compiler. You can also download a copy of the already configured compiler from the Elektor website, saving yourself some work. Skip to the section "Installing the compiler from the Elektor website" if you would prefer.

DOS Box/Windows Console

If you haven't used the DOS box or Windows console before, in Windows click **start → Run...** and then type **`cmd`** in the **Run** dialog box and then press the **`Enter`** key (or click the **OK** button) (**Figure 1.1**). This will start a DOS box (**Figure 1.2**).

You can use DOS commands to navigate through the directories (folders) on the computer, change between drives, run programs, etc. The flashing cursor that you see is known as the "command prompt" (or "DOS prompt") and is where you type the DOS commands. You can close the DOS box for now, it will be used to compile and run our C programs after we have installed the compiler.

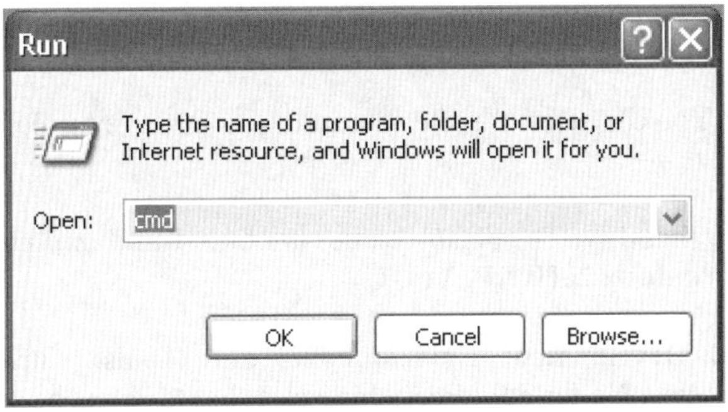

Figure 1.1: Windows "Run" Dialog Box

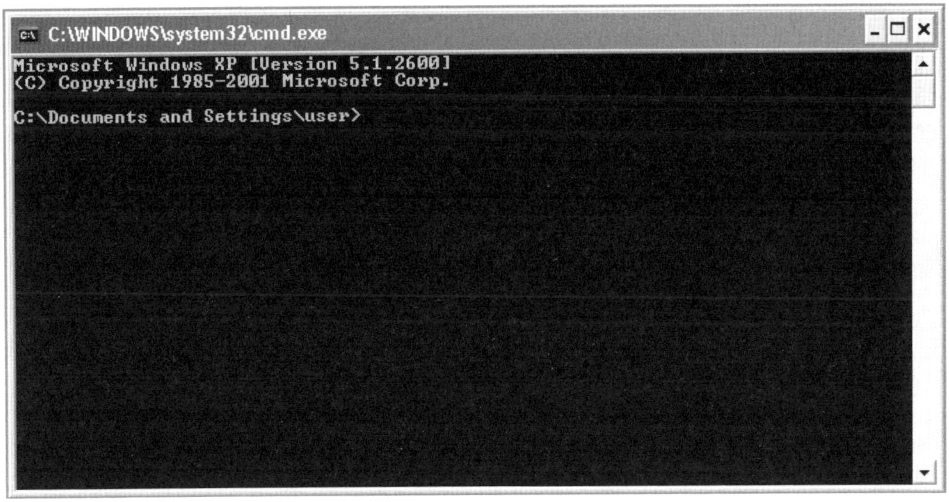

Figure 1.2: Console Window / DOS Box

1.2 Downloading the DJGPP Compiler

These steps are included here in order to show you how to manually download, configure and install the DJGPP compiler. As mentioned, there is a quicker way to download and install the compiler from the Elektor website. The instructions in this section have been included so that if you want to download a newer version of the compiler when it becomes available, you will know what to do. Also if you want to know how the whole download and installation process works, it is described in full detail here. Skip to "Installing the compiler from the Elektor website" if you are not interested in this process.

The DJGPP compiler can be found at the following website: **http://www.delorie.com/djgpp/**. From this web page click the "**Zip Picker**" link. This will take you to **http://www.delorie.com/djgpp/zip-picker.html**. The zip picker page is used to select the components that you will need to install in order to run the DJGPP compiler on Windows.

On the Zip Picker page:

1. **FTP Site:** You can select the site nearest to you or leave the default "**Pick one for me**".

2. **Basic Functionality:** The default values are fine (**Build and run programs with DJGPP**, **Windows 2000/XP**, **Yes**, **C**)

3. **Integrated Development Environments and Tools:** Uncheck the **RHIDE** checkbox, select **No** for **gdb**. Everything else should be unchecked.

4. **Extra Stuff:** Leave everything unchecked.

5. Click the "**Tell me which files I need**" button.

A web page should now appear with the following lines near the middle of the page (you may have to scroll down to see them):

```
unzip32.exe          to unzip the zip files          95 kb

v2/copying.dj        DJGPP Copyright info             3 kb
v2/djdev203.zip      DJGPP Basic Development Kit    1.5 mb
v2/faq230b.zip       Frequently Asked Questions     664 kb
v2/readme.1st        Installation instructions       22 kb

v2gnu/bnu217b.zip    Basic assembler, linker        3.9 mb
v2gnu/gcc423b.zip    Basic GCC compiler             4.3 mb
v2gnu/mak3791b.zip   Make (processes makefiles)     267 kb
v2gnu/txi412b.zip    Info file viewer               1.0 mb
```

Total bytes to download: `12,276,065`

Note that there may be a difference between the version numbers in the file names and the file sizes in this book and the ones on the website.

Save the following items to your computer:

1. DJGPP Basic Development Kit (**djdev203.zip**)
2. Frequently Asked Questions (**faq230b.zip**)
3. Basic assembler, linker (**bnu217b.zip**)
4. Basic GCC compiler (**gcc423b.zip**)
5. Make (processes makefiles) (**mak3791b.zip**)
6. Info file viewer (**txi412b.zip**)

Strictly speaking, you only need to download items 1, 3, 4 and 5 above, but the other packages give you help information and an FAQ document.

1.3 Installing the DJGPP Compiler

DJGPP does not have an installation program that does the install for you. You will need to create a directory and then unzip each package into that directory. I used a program called **7-zip** to do this. You can download it at **http://www.7-zip.org/**. Download and install the latest stable version of **7-zip**, not the beta version. The downloaded file is an executable installation file and will guide you through the installation process.

Do the following to install the DJGPP compiler:

1. Create a directory (folder) on your C drive called "DJGPP" i.e. `C:\DJGPP`
2. Using **7-zip** unzip each of the downloaded packages to the DJGPP directory. (In Windows right click a package, under the **7-zip** menu choose "**Extract files...**" (**Figure 1.3**), click the "**...**" button, navigate to `C:\DJGPP`, click "**OK**" (**Figure 1.4**))

In **figure 1.3** you can see that I have selected all of the DJGPP packages that were downloaded and right clicked on them to bring up the **7-zip** menu. You can unzip all of the packages simultaneously like this or do each one individually.

Figure 1.4 shows **7-zip's Extract** dialog box, click the "**...**" button to navigate to the DJGPP directory and then click the "**OK**" button to unzip the packages there.

The **C_Tools** directory shown in **figure 1.3** contains all of the files that you will download in this chapter. Note that the version numbers in the file names may differ from the ones that you have downloaded.

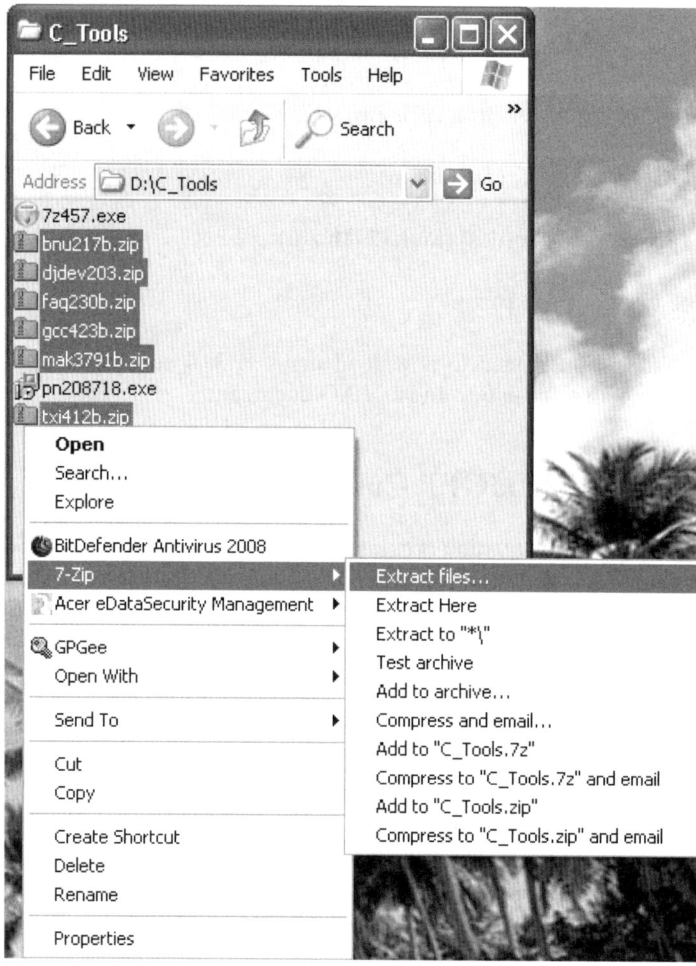

Figure 1.3: Using 7-zip to Extract the DJGPP Compiler Tools

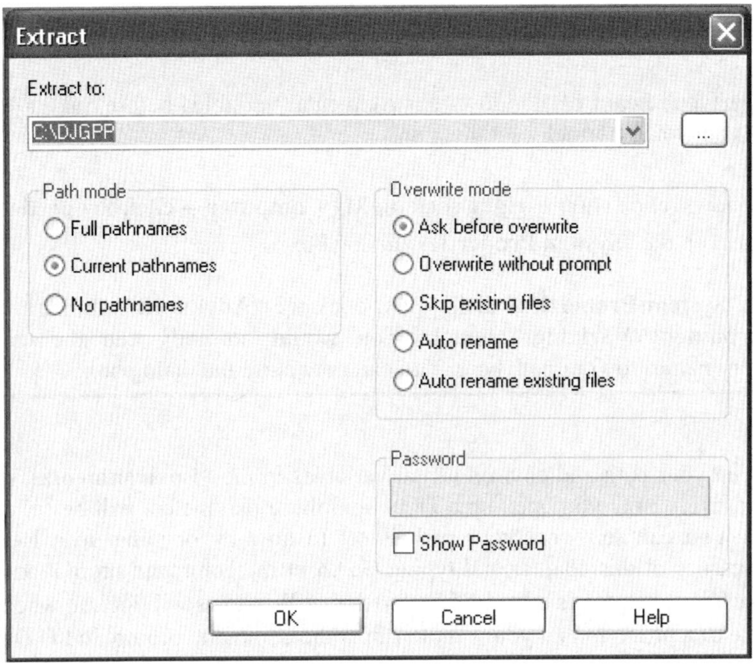

Figure 1.4: 7-zip Extract Dialog Box

1.4 Setting Up the DJGPP Compiler

First you will need to create a directory where you will keep the programs that you are going to write. My hard drive is partitioned into two drives C: and D:. I like to store programs on the C: drive and my data and work on the D: drive, so I will make a directory called **`cprogs`** on the D: drive i.e. **`D:\cprogs`**. You can make your directory on any drive that you want and call it anything that you want. Don't use spaces in your directory name, and keep it short as you will have to type it in at the DOS prompt and a short name will save you some typing.

The compiler will not be able to run yet because it needs some parameters to be set up first. This information is available in the **Readme** file of the compiler directory (**`C:\DJGPP\readme.1st`**). You can read this file by opening it with any text editor e.g. **Notepad** in Windows. The setup requires that you must add certain environmental variables to Windows as well as add some directory names to your Windows "path". Have a look at the **Readme** file if you like, but I will explain how to set the compiler up.

> **Windows Environmental Variables**
>
> If you haven't heard of Windows environmental variables before, take a look at where you would normally set them up:
>
> In Windows, click **start → right click on My Computer → click Properties**. This will open up the "**System Properties**" dialog box.
>
> In the "**System Properties**" dialog box, click the "**Advanced**" tab and then the "**Environment Variables**" button. You would normally add the compiler directory names to your path here. You can now close the dialog box.

If you set up your path and environmental variables in the "**Environmental Variables**" dialog box, every time you open up a DOS box, these parameters will be set up for that DOS box. You can see what your path is set to now by opening up a DOS box as described earlier in this chapter and typing **`path`** at the command prompt and pressing the **`Enter`** key. The path is a list of directories that Windows will look in when you type in the name of a program that you want to run at the command prompt in the DOS box. If the directory of your compiler is not present in the path, Windows will not be able to find it and run the compiler; instead it will display an error message.

The problem with setting up your path globally in Windows (i.e. using the **Environment Variables** dialog box) is that if you want to set up more than one compiler, the compilers may end up clashing. For example if two compilers have utilities with the same name, the one that appears first in the Windows path will be the one that is run every time. To solve this problem, we will use batch files.

Another reason for using batch files is that it is much easier to remove the compiler from your system without having to remember what you changed in your Windows path. If you ever want to remove the compiler, just delete the directory that it resides in with all of its contents.

We will start by creating a batch file to set up the DJGPP environment. A batch file is just a text file that has a **.bat** file extension instead of a **.txt** file extension. It contains DOS commands that will run when you double click it in Windows.

In Windows, navigate to the DJGPP compiler directory (**`C:\DJGPP`**) using **start → My Computer**, right click in the DJGPP directory (not on one of the icons in the directory, but on the background of the directory) and select **New → Text Document** from the pop-up menu. Name the file **`djgppbox.bat`**. When Windows asks if you want to change the filename extension, click "**Yes**". If you can't see or change the filename extension, you need to enable this in Windows. To do this, navigate to the DJGPP directory using

start → My Computer if you haven't already done so. Click the "**Tools**" menu, click "**Folder Options...**", click the "**View**" tab in the "**Folder Options**" dialog box, uncheck the "**Hide extensions for known file types**" check box, click the "**OK**" button.

You will now edit the **djgppbox.bat** file that you have made. Right click the file and click "**Edit**" from the pop-up menu. The file should open in **Notepad**.

Add the following lines to the file:

```
@echo off
set PATH=c:\djgpp\bin;%PATH%
set DJGPP=c:\djgpp\djgpp.env
D:
cd D:\cprogs
cmd
```

Lines 4 and 5 of the listing above are used to change to your directory that you created earlier in order to store your C programs. If you have chosen to name your directory differently to mine, then you will need to change that in lines 4 and 5. E.g. if your directory is named "**myprogs**" and is on the **C:** drive, change line 4 to **C:** and line 5 to **cd C:\myprogs**. Save the batch file that you have created and close **Notepad**.

You can now double click the **djgppbox.bat** icon and it will open up a DOS box with the environment set up for the DJGPP compiler. The DOS box will open in the directory that you created to store your program files in. To test that the compiler is installed correctly, type **gcc --version** at the DOS prompt. You should get back a version number and copyright information (**Figure 1.5**). Close the DOS box for now.

If the compiler did not work, go back and check that you set everything up correctly and that there are no typing errors in the **djgppbox.bat** batch file.

An Explanation of the djgppbox.bat Batch File

Line1: Stops the rest of the lines in the batch file from being displayed on the screen when the batch file is run.

Line 2: Prepends the directory in which the compiler resides to the Windows path. To check this type **path** in a DOS box that is started with **djgppbox.bat** and you will see that the path starts with **PATH=c:\djgpp\bin;...**

Line 3: An environmental variable that the compiler requires.

Line 4: Changes drives to where your chosen programming directory was created, in this case to the D: drive.

Line 5: Changes to the chosen programming directory.

Line 6: Starts the DOS box.

1.5 Installing the Compiler from the Elektor Website

If you have already installed the compiler manually, you do not need to follow these steps, you can move to the next section – "How C Programs are Created".

Go to the website **http://www.elektor.com** and download the file **DJGPP.zip** Extract the contents of the zip file to your C: directory. You can open the zip file by double clicking it in Windows. It contains the directory DJGPP, drag and drop this directory onto your C: drive i.e. **C:\DJGPP**.

You now need to create a directory in which you will store the programs that you are going to write. You will also need to edit a batch file to point to this directory. The batch file currently points to a directory called **cprogs** on the **D:** drive. If you have a **D:** drive and want to make a directory there called **cprogs**, to store your C programs, you can create this directory and then you will have finished setting up the compiler. You will just need to test it as shown below.

If you do not have a D: drive and/or want to create a directory with a different name, do so now. You will need to edit the batch file in the DJGPP directory i.e navigate to the DJGPP directory in Windows **C:\DJGPP**, right click the file **djgppbox.bat** and then click "**Edit**" on the pop-up menu. The file should open in Windows **Notepad**.

Change the line **D:** to reflect the drive that contains your programming directory. E.g. if you made the directory on the C: drive, change **D:** to **C:** in the file.

The line with the statement **`cd D:\cprogs`** must be changed to reflect the directory that you have chosen. E.g. if you decided to make a directory call **`myprogs`** on the **`C:`** drive, then you must change this line to **`C:\myprogs`**. Save the file. You are now ready to test the compiler.

In Windows, navigate to the directory containing the DJGPP compiler i.e. **`C:\DJGPP`**. Double click the **`djgppbox.bat`** icon to start a DOS box. Type **`gcc --version`** at the command prompt and then press the **`Enter`** key. You should get back version and copyright information as shown in **figure 1.5**.

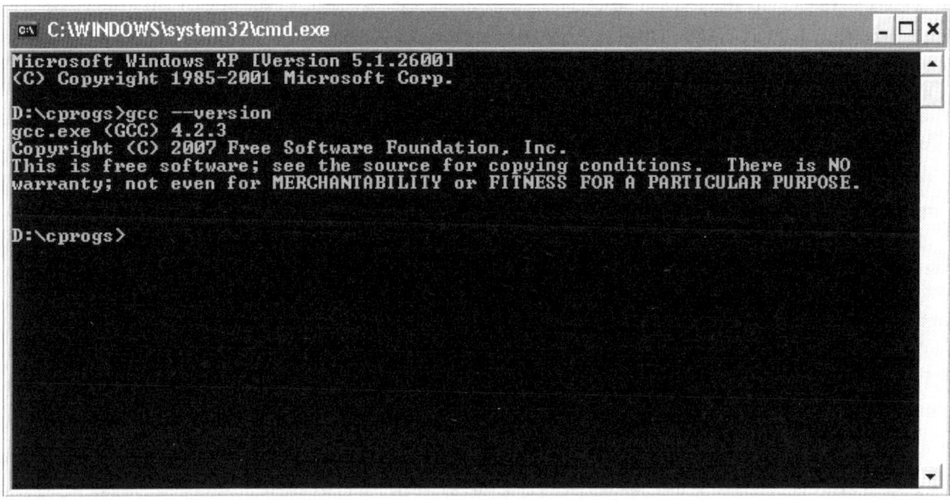

Figure 1.5: Testing the DJGPP C Compiler

Now that your compiler is installed and working, we can look at how C programs are written and install a program to help us create and edit C programs.

1.6 How C Programs are Created

C programs are written using a text editor and saved to a file with a **.c** file extension e.g. **`myprog.c`**. When the C compiler is run it reads the C text file, known as the C source file, and creates an executable file that can run on the PC. You could use **Notepad** in Windows to create your C source file (just save the file with a **.c** extension) or any other ASCII text editor, but we will use **Programmer's Notepad** from **http://www.pnotepad.org/**.

Programmer's Notepad has "syntax highlighting" which will highlight C language keywords making your program a lot easier to read. You can not use a word processor for creating C source files because it will add formatting information to the source file (such as font size, bold font, etc.) that will confuse the C compiler.

Go to **http://www.pnotepad.org/** and click the "**Download**" link. On the download page, download the latest stable version of Programmer's Notepad and install it on your PC. Just run the downloaded file and it will take you through the installation process.

1.7 Start Programming

With the software tools installed, you can now write your first C program. Start by creating a new subdirectory in which to store your first C program. This will be a subdirectory of the directory that you created earlier when setting up the compiler (e.g. **D:\cprogs**). You may want to create a subdirectory for each chapter of this book or for each program in this book. I will create a subdirectory called CH1 (chapter 1) to save my first program in (i.e. **D:\cprogs\CH1**).

Create a new C source file: In Programmer's Notepad, click **File → New → C/C++.**

Enter the following lines of code into the new file:

```
#include <stdio.h>

int main(void)
{
    printf("Hello, world!\n");
    return 0;
}
```

Be careful not to leave anything out when typing the program and be sure to type everything exactly as it is shown in the text. Press the **Tab** key to move text to the right for the lines starting with **printf** and **return** (Programmer's Notepad will actually tab the line starting with **return** for you if you have tabbed the **printf** line already).

Save the file as **hello.c** in your chosen directory by typing **hello.c** in the "**Filename**" box of the "**Save As**" dialog box. Your new file should look like **figure 1.6**.

You will have noticed when typing the program into Programmer's Notepad that certain words change colour. This is known as syntax highlighting and makes your program easier to read by highlighting keywords, data types and other elements of the C language.

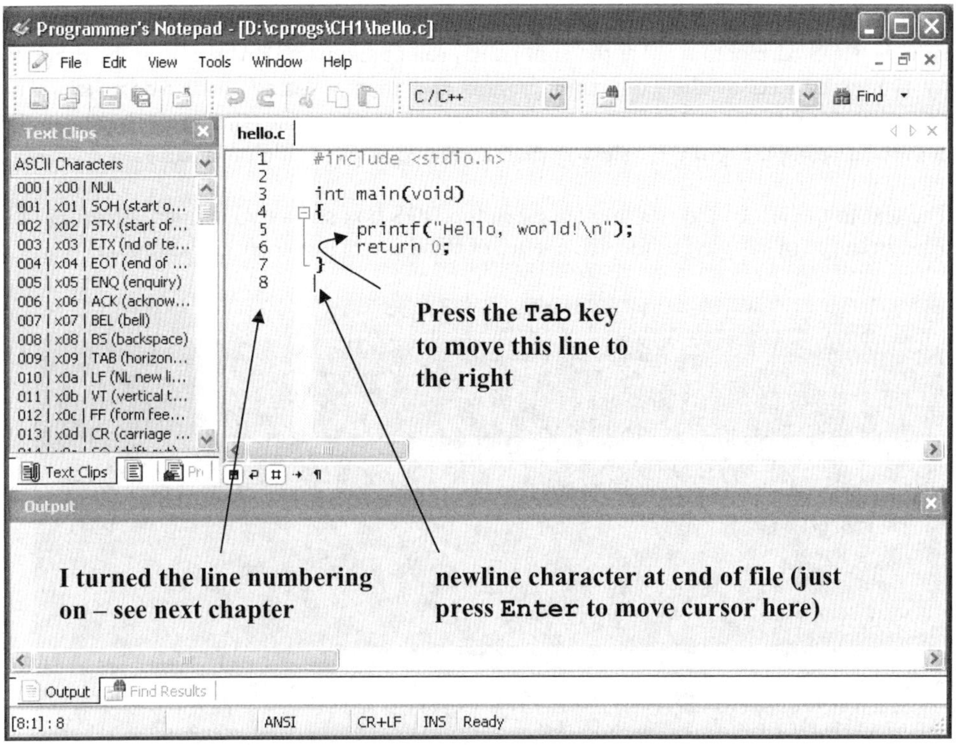

Figure 1.6: Creating the "Hello, world" Program in Programmer's Notepad

1.8 Compiling the Program

To compile the program using DJGPP, navigate to your DJGPP directory and double click the **`djgppbox.bat`** batch file to start a DOS box. In the DOS box, change to the subdirectory that you saved your **`hello.c`** program in using the DOS change directory command (**`cd`**). e.g. **`cd ch1`** then press the **`Enter`** key.

Type the following and then press the **`Enter`** key:

```
gcc hello.c -o hello.exe
```

This line runs the DJGPP **`gcc`** compiler, tells it to compile the **`hello.c`** file and to name the executable file that it creates **`hello.exe`**.

The DJGPP compiler will print a warning message to the screen if there is not a newline character at the end of your C source file, it will still compile the file if there are no other errors. To get rid of the warning, edit your file in Programmer's Notepad by putting the cursor at the end of the closing brace (`}`) and then pressing the **`Enter`** key to move to the next line. This will add a newline character to the end of the file. Save the file and then compile again.

You should now have a new file in your current directory called **`hello.exe`**. To check, type the DOS **`dir`** command at the command prompt and press **`Enter`**. Run the program by typing **`hello`** at the command prompt and then pressing the **`Enter`** key. The program should print **`Hello, world!`** to the screen and then exit back to the command prompt. **Figure 1.7** shows what the above steps should produce.

The text to the left of the flashing cursor in the DOS box shows you which directory you are currently "in" as well as the drive letter of the drive that you are operating on.

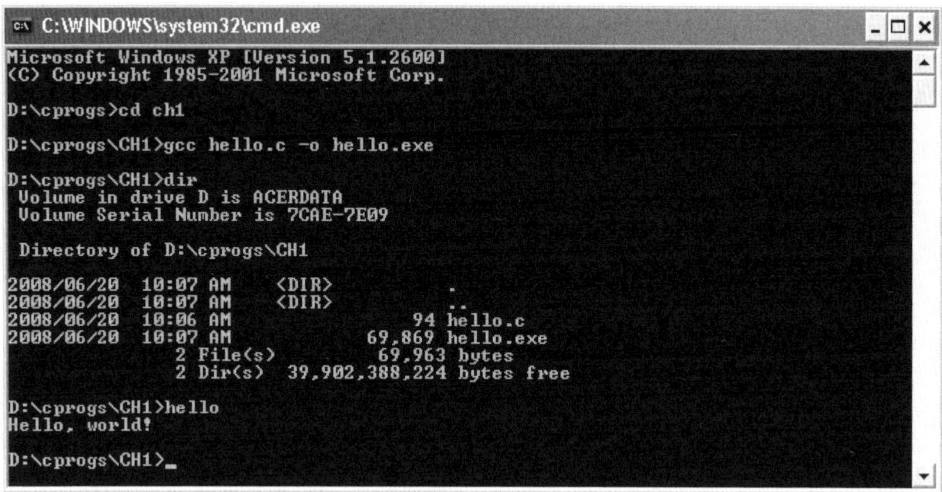

Figure 1.7: Using the DJGPP Compiler

If you had problems compiling the program, here is a summary of all the things that must be in place in order to compile and run the program successfully:

1. Your **`djgppbox.bat`** file must contain the lines as described previously. You should have already tested it previously when doing the setup by typing **`gcc --version`** at the DOS prompt.

2. After starting the DOS box with the **`djgppbox.bat`** file, you must change to the directory or subdirectory that contains the **`hello.c`** program that you saved.

3. Make sure that you have *saved* the file and not just typed it into the editor. Also check that you saved it as **`hello.c`** and not anything else.

4. The **`hello.c`** program must be typed in exactly as it appears in the text, don't leave anything out or add anything. All of the quotation marks, semicolons and brackets must be there and in the correct place. If there were any errors in the C source file, the compiler would have printed error messages without creating

the .exe file. Remember to save the file before compiling again if you have to make any corrections.

5. The line that you type in at the command prompt to run the compiler must be typed in exactly as it appears in the text above.

All of the source code files used in this book are available for download from the Elektor website. If your program still does not work, you can try to compile the **`hello.c`** file from Elektor. Refer to appendix B for a listing of the source files used in this book.

If you try to run the **`hello`** program by double clicking the **`hello.exe`** icon in Windows, a DOS box will come up and then quickly disappear. What is happening here is that the program opens in a DOS box, runs, and when finished Windows closes the DOS box automatically. We will solve this problem in the next chapter.

You can navigate to the **`hello.exe`** file to double click it using **My Computer** or type **`explorer .`** and press **`Enter`** at the command prompt to open up the current directory in a window. Don't forget to type the dot after typing explorer.

Some Useful DOS Commands

If you don't know any DOS commands, here are some that you may find useful. Each command is typed in at the DOS prompt in the DOS box and then the **`Enter`** key is pressed to run the command. DOS commands are not case-sensitive, so **`CD`** and **`cd`** are the same to DOS. These commands can also be used in batch files.

Changing Directories:
Command: **`cd`**
Examples:
`cd programs` changes to the **`programs`** directory if it exists. You can only change to a directory that is a subdirectory of the one that you are currently in unless you type the full path as shown below.
`cd programs\ch1` changes to the **`ch1`** subdirectory in the `programs` directory.
`cd ..` moves back one directory from where you are currently.
`cd \` moves to the root directory of the drive.
`cd C:\programs\ch1` specifies the full path of the directory to change to. You can leave out the drive letter if you like (**`C:`**).

Change Between Drives:
Just type the drive letter followed by a colon. e.g. **`C:`** to change to the C drive, **`D:`** to change to the D drive, etc.

Clear the Screen:
`cls`

Make a Directory:
Command: **`md`**
Example: **`md CH2`** makes a directory called **`CH2`** in the current directory.

Running a Program:
Just type the name of the program. e.g. **`hello`** to run the **`hello.exe`** program that you created or **`notepad`** to launch Windows Notepad from the DOS box. You must be "in" the directory in which the program exists i.e. changed to that directory using the DOS **`cd`** command, or the directory must be in the Windows path (as in the case of running Notepad from the DOS box).

Pressing the up arrow key will show the previous DOS command that you typed. Pressing it again will show the command that you typed before that and so on for each command that you previously typed. Pressing the down arrow key will scroll back down through the list. Pressing **`Enter`** will run the DOS command that you scroll to, saving you some typing.

Everything else you will probably want to do using Windows, like deleting or moving files etc.

1.9 About Your First C Program

The program that you have written is a classical first program for any computer language – printing "Hello, world!" to the computer screen.

The first thing that you should know about C is that it is case sensitive. If you change the case of any of the keywords or function names in your program and try to compile it, you will get an error message. In the **`hello.c`** program **`printf`** is a function name and **`return`** is a key word. If you typed **`Printf`** or **`Return`** or **RETURN** for example, the compiler will display an error message and will not be able to compile the program.

Try changing the case of the keyword or function name, then try to compile if you would like to experiment. Don't forget to save your source code file before compiling. Note that if you successfully compiled your program previously, the **`hello.exe`** file will already exist. If there are now errors in your source code, the compiler will not overwrite or erase the already existing **`hello.exe`** file. It won't produce any output except for the error message. Don't be fooled into thinking that it created a working **`hello.exe`** file. To prevent this, delete the old **`hello.exe`** file and then you can be sure if the compiler has or has not produced a new **`hello.exe`** file when compiling again.

The text that is printed to the screen between the quotation marks (`""`) can be changed as this is just a message and not part of the C language. Type any message between the quotation marks that you would like and then save, compile and run the program.

The **`\n`** that forms part of the text between quotation marks moves the cursor to a new line. It tells the compiler to insert a newline character at the end of the text message. Try adding some more **`\n`**'s to the text, save, compile and run. e.g. **`printf("Hello, world!\n\n\n\n");`** will move the cursor down four times on the screen so that when the program is run, there will be a gap between the command prompt and the text printed by the **`hello.exe`** program. You can also try putting a newline character in front of the text or in the middle of the text and see what happens. Just make sure that it is between the quotation marks.

1.10 Analysing the Program

We will now examine each element of the **`hello.c`** program. Refer to **figure 1.8** when reading this section.

1.10.1 Characters and Strings

A character is any keyboard character that you can type such as letters of the alphabet, numbers and special characters like $, %, !, etc. A string is a sequence of characters – written sequentially in a program and stored sequentially in the computer's memory. In C, strings are written between quotation marks as in "Hello, world!\n".

When you press the **Enter** key on the keyboard, you are typing a newline character. The newline character moves the cursor to a new line as we saw in our program. If you want to insert a newline character into a text string when editing your C source file, and you try to do so by pressing the **Enter** key, the cursor in your text editor program will move to the next line in the source file. This is because the text editor is interpreting the newline character that you type. To insert a newline character into your text string for your program to use, we use an "escape sequence" – this is the **\n** that you have already seen. There are other escape sequences besides the newline, they all start with a backslash (\) when written in C programs.

1.10.2 Functions

A function in C is a piece of code that can be called (run) from within a C program to perform some task (or function). In the text a function is written with the function name first and then two parentheses following it, e.g. **myfunc()**. Functions are called subroutines in some programming languages.

Every C program has to have a function called **main()**. This is the entry point of the C program – where it will start running. The **main()** function has an opening and closing brace (**{** and **}**) around the code that it contains. This lets the compiler know that the code within the braces belongs to the **main()** function. Ignore the **int** and **void** keywords at the **main()** function for now.

The **printf()** function is an external function that is called by the **main()** function in the **hello** program. What this means is that by typing the function name **printf()** in your C source code, a function called **printf()** will be run when the **hello** program is run. The **printf()** function exists somewhere externally and gets added to your program when the compiler tools are run. All you need to know for now is that the **printf()** function will print text to the screen when a line of text is passed to it between quotation marks.

This is all that you need to know about functions for now; they will be covered in more detail later in the book.

1.10.3 Program Statements

The lines of code between the braces of the **main()** function are each examples of C statements. Note that every statement in C must be terminated with a semicolon (;). If the semicolon is left out, the compiler will display an error message and will not compile the source code.
Ignore the **return** statement for now. Just know that the compiler will complain if it is not there.

1.10.4 Preprocessor Directives

Line 1 of the **`hello.c`** program contains a preprocessor directive: **`#include <stdio.h>`**. This is telling the compiler to include an external file named **`stdio.h`**. For now all you have to know about this line is that it must be present at the top of your source code file if you want to use the **`printf()`** function.

1.10.5 Whitespace Characters

The C compiler will ignore all whitespace characters in the source code. Whitespace characters are spaces, tabs and the newline character. So you could separate each line of the **`hello`** program vertically by hundreds of newline characters. Of course this would make the program difficult to read, but the compiler would still compile the program.

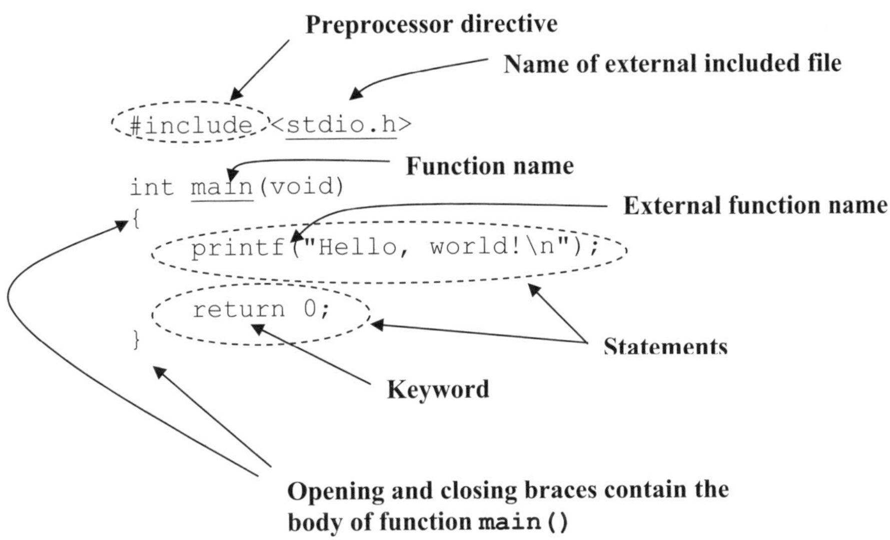

Figure 1.8: Elements of a C Program

1.11 Exercises

The best way of learning to program is to write programs. The following exercises are very simple, but be sure to complete them in order to reaffirm what you have learned in this chapter. The solutions to the exercises follow immediately after, but don't look at them until you have attempted to do the exercises yourself. I will suggest that you make a subdirectory in your programming directory called **`exercises`** in which to save your programming exercises. Appendix B contains information on downloading the source code programs and exercises used in this book.

1. Write a program that uses a single **printf()** function to write three lines of text to the screen. The output should look like this:

```
The C
language
is great!
```

2. The following program contains three errors, see if you can spot them. If you can't, then create the program and try to compile it. The error messages from the compiler should help you to find and correct the errors.

```
#include <stdio.h>

int main(void)
{
    printf("I can use two printf()\n")
    printf("statements in my program.")
    return 0
}
```

1.11.1 Solutions

1.
```
#include <stdio.h>
int main(void)
{
    printf("The C\nlanguage\nis great!\n");
    return 0;
}
```

2. The terminating semicolon has been left off the end of each of the statements in the program.

1.12 Summary

- Command line tools need environmental variables to be set up including the path in order to work.

- C programs are written with a text editor and saved in a text file that ends with a **.c** file extension. This file is known as a source file and contains source code.

- When the C compiler is run, it produces an executable file or computer program. (The tools that run actually compile and then link – more about this later).

- C is case sensitive. Functions and keywords must be written in the correct case. C uses mostly lowercase for keywords and function names.

- Every C program must start with a **`main()`** function. This is where the program starts running.

- The **`main()`** function and every other function that you write must have it's code placed between braces **`{ }`**.

- All function names end with parentheses **`()`**.

- Each statement in C must be terminated with a semicolon `;`.

- Common typing errors when beginning with the C programming language are: leaving off the semicolon at the end of a statement, leaving off one or both of the enclosing braces of a function and leaving off the parentheses of a function.

Congratulations, you have come a long way in this chapter! We will move on to some more programming in the next chapter and set up Programmer's Notepad to make the compiling and linking process easier. If you do not understand everything in this chapter, do not worry. Things will become clearer as you progress through the book.

2 C Basics

What you will do in this chapter:

- Set up Programmer's Notepad to make compiling easier
- Write some more simple C programs
- Learn about input, output and variables
- Do some simple mathematical calculations
- Learn how to find and fix program errors

There are a number of example programs in this chapter and the chapters that follow. It is advisable to enter and run all the example programs in this book. Feel free to change the programs and experiment with them once you understand how they work. This will help you to learn C.

2.1 Setting up Programmer's Notepad

Commercial programming tools will typically include an IDE (Integrated Development Environment). This is a program that includes an editor for editing source code, usually a way of saving your work as a project, a debugger to help find programming errors (bugs) and the ability to compile and run the programs that you write all from within the IDE. It saves you from having to compile and run your programs from the command line; everything that you do is done through the IDE.

It is easy to work with an IDE, everything is there for you and should work together nicely with minimal or no extra setup required after installing. The object of setting up the compiler manually in the previous chapter and in this one is to teach you what to do if a compiler that you want to use does not come packaged nicely with an IDE.

We can set up Programmer's Notepad to act like an IDE. You will be able to compile your source code from within Programmer's Notepad. Before we set Programmer's Notepad up to run the compiler, let's change a few settings in Programmer's Notepad to make it easier to use.

2.1.1 Some General Settings

These settings are optional, I will run you through the settings that I prefer and you can decide if you like them or not. I will give you an explanation of why I use some of the settings to help you decide if you want to use them yourself.

1. Start Programmer's Notepad.

2. Click the **Tools** menu and then **Options**.

3. Click the **General** item in the left pane of the **Options** dialog box to highlight it. (See **figure 2.1** for steps 3 to 8)

4. Look at the check boxes to the right of the pane.

5. Check "**Maximise new document windows**" by clicking it. This will maximise any new document that you create within Programmer's Notepad. I find that I do this anyway when I work with a new document, so prefer to have it set as the default.

6. Check "**Show full file path in document windows**". This is handy if you have two files open that have the same file name. Displaying the path (where they are located on the hard drive), helps to determine which one is which.

7. Uncheck the "**Start with a new, blank document**". I don't always open Programmer's Notepad to start a new document, sometimes only to open and edit an already existing one, so a new one has to be closed anyway.

8. The "**Remember open files and projects between sessions**" may be useful later as it will always open up your last edited file after you have closed and then re-started Programmer's Notepad.

9. In the left pane, click **Defaults** under the **General** item. (See **figure 2.2** for steps 9 to 11)

10. Checking "**Show line Numbers**" will show the number of each line in a source code file. This is very handy when searching for compile errors. The compiler will tell you which line number the error is on. Having the line numbers displayed on the left will help you to find the line number quickly.

11. Uncheck "**Use the tab character for indentation**". This will put spaces into your source file when you press the `Tab` key instead of the tab character. If you use spaces instead of tabs, your source code will display correctly no matter which editor you open it in. This is because the tab character can be set up to different widths in different editors. The default tab width is set up as 4, which is fine. (It will now use 4 spaces in place of inserting a tab character)

12. Click **Schemes** in the left pane.

13. I like to change the background colour here from the default white. A white background can result in eye strain and headaches when programming for a long time. Change the background colour if you like, but don't change it to a solid colour that is used as one of the syntax highlighting colours. If you do, the text that uses that colour will be invisible. I prefer to choose a light shade of grey, green or blue.

14. Click the **OK** button to finish.

You can close the **Scripts** box that is at the right of Programmer's Notepad if you would like more screen space for editing your documents.

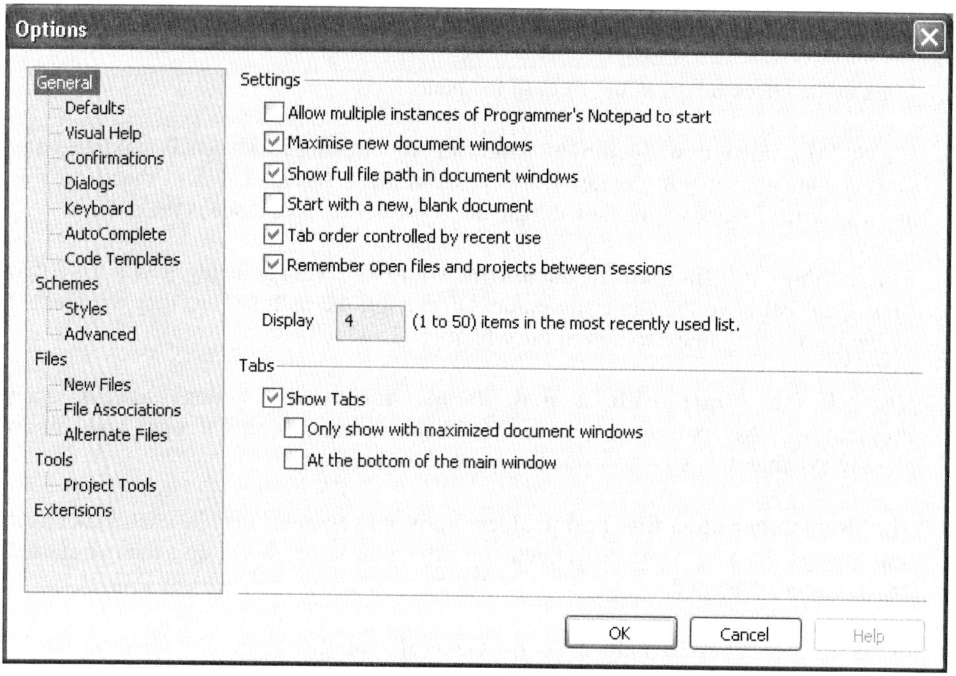

Figure 2.1: Programmer's Notepad Setup Steps 3 to 8

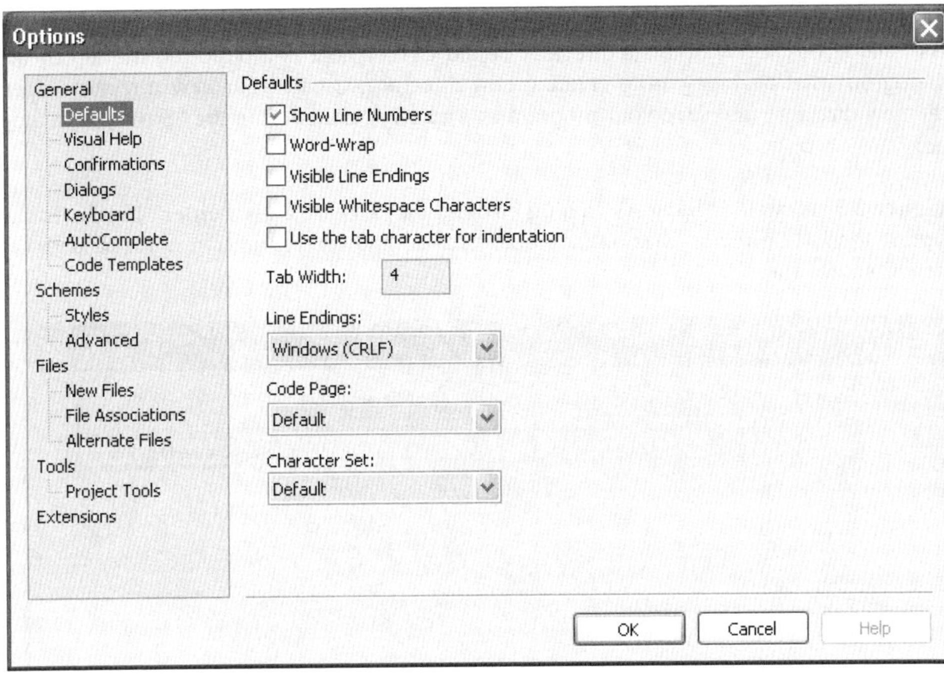

Figure 2.2: Programmer's Notepad Setup Steps 9 to 11

2.1.2 A New Program to Compile

We will now create a new program so that you have something to compile when we set up Programmer's Notepad to run the compiler.

Start Programmer's Notepad, select **File** → **New** → **C/C++**. You could also click the **New** icon, but selecting **C/C++** enables the syntax highlighting immediately – Programmer's Notepad does not know what language you will be programming in if you select the **New** icon as it supports multiple programming languages. Enter the following program into the new file:

```
#include <stdio.h>
#include <conio.h>

int main(void)
{
    printf("The console window will stay open.\n\n");
    printf("Press any key to close the window.");
    getch();

    return 0;
}
```

When you save the file in Programmer's Notepad, use the "**Save As**" dialog box to navigate to the programming directory that you created earlier. In the previous chapter, I decided to save my programs to a subdirectory for each chapter, so I will use the "**Save As**" dialog box to create a subdirectory called CH2. There is a button at the top of the dialog box that enables you to create a new directory. Create your new directory, open the new directory and save your program as **`sticky.c`** all within the "**Save As**" dialog box.

Figure 2.3 shows the "**Save As**" dialog box and the "**Create New Folder**" button. In the figure, the CH2 directory has already been created and entered into in order to save the new C source file.

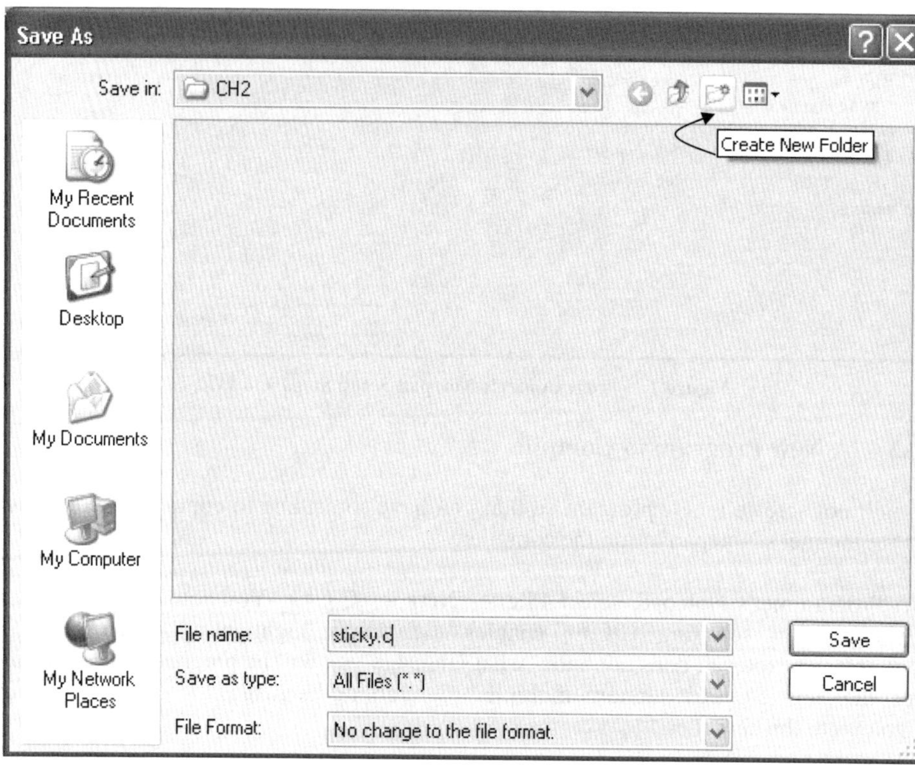

Figure 2.3: Programmer's Notepad Save As Dialog Box

Programmer's Notepad has the ability to run external tools like our compiler. Because we haven't set up our tools environmental variables so that every program can "see" them i.e. globally, we will have to set up a batch file for the compiler again. If you downloaded the compiler from the Elektor website, this batch file has already been created for you, but you will still have to set up Programmer's Notepad to use it.

The next section shows you how to create the batch file and then set up Programmer's Notepad. Skip the batch file creation if you installed your compiler from the Elektor website. You will need to do the Programmer's Notepad setup though.

2.1.3 Setting up Programmer's Notepad to Run the DJGPP Compiler

Batch file setup:

1. Create a new batch file in the DJGPP compiler directory i.e. **C:\DJGPP**, as done in chapter 1.

2. Name the file **`djenv.bat`**.

3. Open the file for editing and add the following lines (You can copy the first three lines from **`djgppbox.bat`**):

```
@echo off
set PATH=c:\djgpp\bin;%PATH%
set DJGPP=c:\djgpp\djgpp.env
gcc %1 %2 %3
```

4. Save and close the file.

Programmer's Notepad setup:

1. In Programmer's Notepad, click **Tools** → **Options**.

2. In the pane on the left inside the **Options** dialog box, click **Tools**.

3. If you do not have a C source file open in Programmer's Notepad with the C source file being the currently selected file, you will need to select **C/C++** in the **Scheme:** drop down list box.

4. Click the **Add** button.

5. In the **Name:** box, type: **`DJGPP Compiler`**

6. Use the "**...**" button on the right of the **Command:** box to navigate to the DJGPP compiler directory (**`C:\DJGPP`**). Click the **`djenv.bat`** batch file to select it and then click the **Open** button.

7. Under **Folder:** type: **`%d`**

8. In the **Parameters:** box, type: **`%f -o %n.exe`**
9. Make sure that the "**This tool will modify the current file**" box is unchecked.

10. **Figure 2.4** shows what you should have achieved so far.

11. Click the **OK** button.

12. Click the **OK** button in the **Options** dialog box.

Under **Tools** in Programmer's Notepad, you should now have a menu item named "**DJGPP Compiler**". Click it to compile the **`sticky.c`** file using the DJGPP compiler. The **`sticky.c`** file must be open in Programmer's Notepad and must be the current file being viewed in order for this to work. The output messages from the compiler will now be displayed in the **Output** window in Programmer's Notepad.

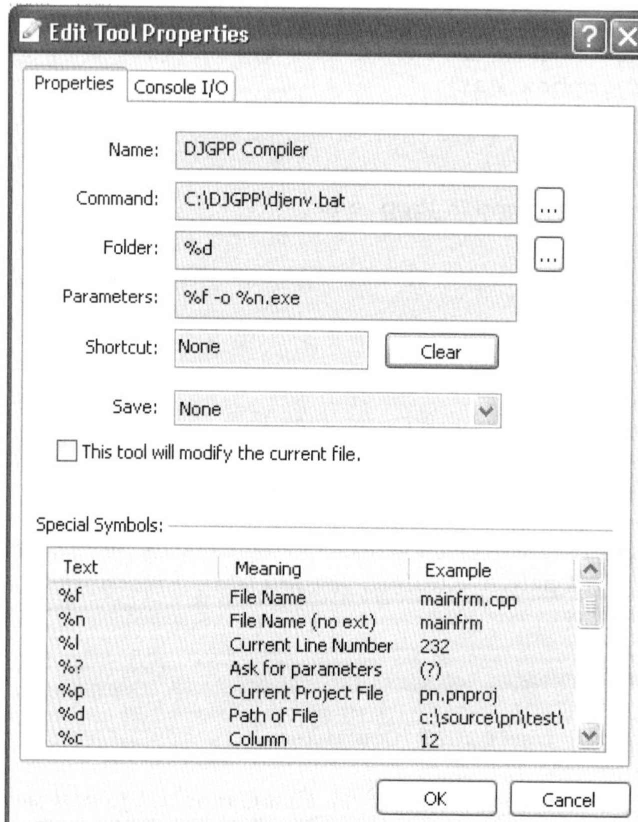

Figure 2.4: Tools Setup in Programmer's Notepad

Run the **`sticky.exe`** program that is created from the compile by double clicking it in Windows. You will see that the DOS box does not disappear like it did with the **`hello`** program in chapter 1.

You have now set up your compiler to run under Programmer's Notepad. The program that you have written can also be run from Windows without having to run it manually from the command line, so you can now compile and run your programs all from within Windows.

Figure 2.5 shows what you should have achieved. The figure shows the source file and output from the compile to the right of Programmer's Notepad in a window opened from "**My Computer**" so that **`sticky.exe`** can be double clicked to run it.

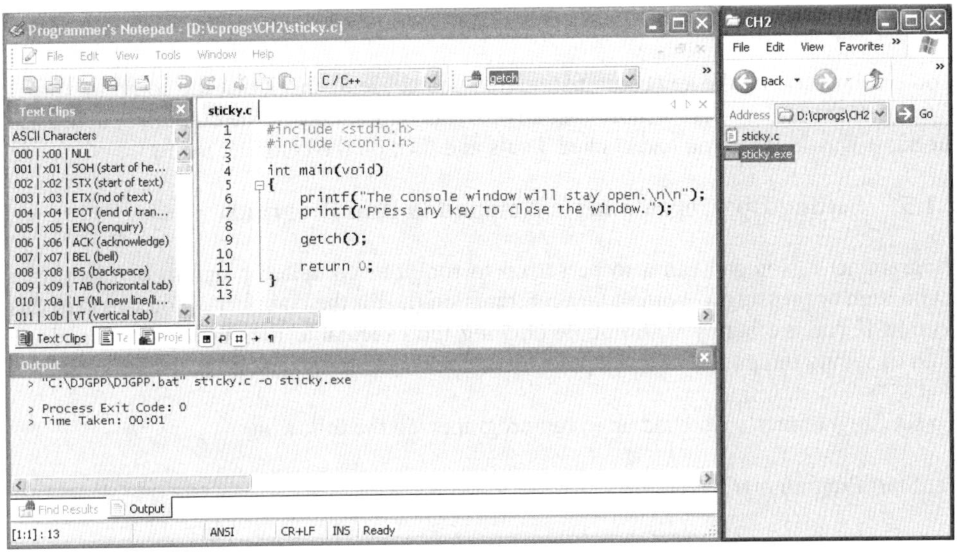

Figure 2.5: Running DJGPP from Programmer's Notepad

2.1.4 A Keyboard Shortcut for the Compiler

One last setting that you might like to do is to add a shortcut key for the compiler. This will enable you to use a key combination on the keyboard to do the compile instead of having to click the Tools menu and then on the compiler name.

To add a shortcut key do the following:

1. Click **Tools** and then **Options** in Programmer's Notepad.

2. Click **Tools** in the left pane of the **Options** dialog box.

3. Select **C/C++** in the **Scheme:** drop down list. This is only necessary if you do not have a C source file open and selected in Programmer's Notepad.

4. Click on the compiler name that you would like to add the keyboard shortcut to. This should be **`DJGPP Compiler`** if you chose that name when setting up Programmer's Notepad previously.

5. Click the **Edit** button.

6. Click in the **Shortcut:** box to move the cursor there.

7. Type the key combination that you would like to use for the compiler. I am going to choose F9 for the DJGPP compiler. I.e. the F9 function key at the top of the keyboard.

8. Click the **OK** button and then the **OK** button in the **Options** dialog box when finished.

You can now try out the key combination that you chose to check that the compiler runs. E.g. Press the F9 key if you chose that key as your shortcut key. The compiler should run the same way it did if you had clicked **Tools** and then on the compiler name.

2.1.5 Setting Up Programmer's Notepad to Run Programs

Programmer's Notepad can also be set up to run your compiled program by clicking a menu item or pressing a shortcut key combination (as for the compiler). You can skip this section if you are happy with double-clicking the executable files in Windows and are tired of setting things up. This setup however does make things easier.

To set Programmer's Notepad up to run programs, do the following:

1. Start Programmer's Notepad.

2. Click **Tools** and then **Options**.

3. Click **Tools** in the left pane of the **Options** dialog box.

4. Select **C/C++** in the **Scheme:** drop down list box.

5. Click the **Add** button.

6. In the **Name:** box type: `Run exe`
   (See **figure 2.6**)

7. In the **Command:** box type: `%n`

8. In the **Folder:** box type: `%d`

9. Leave **Parameters:** empty.

10. Click in the **Shortcut:** box to move the cursor there and press the keyboard shortcut key combination that you would like to use to run your compiled programs. I will choose to use the F10 key.

11. The **Save:** box must contain: `None`

12. The "**This tool will modify the current file.**" checkbox must be unchecked.

13. Click the "**Console I/O**" tab at the top of the "**Edit Tool Properties**" dialog box. (See **figure 2.7**)

14. Uncheck the "**Capture Output**" checkbox.

15. Click **OK**.

16. Click **OK** to close the **Options** dialog box.

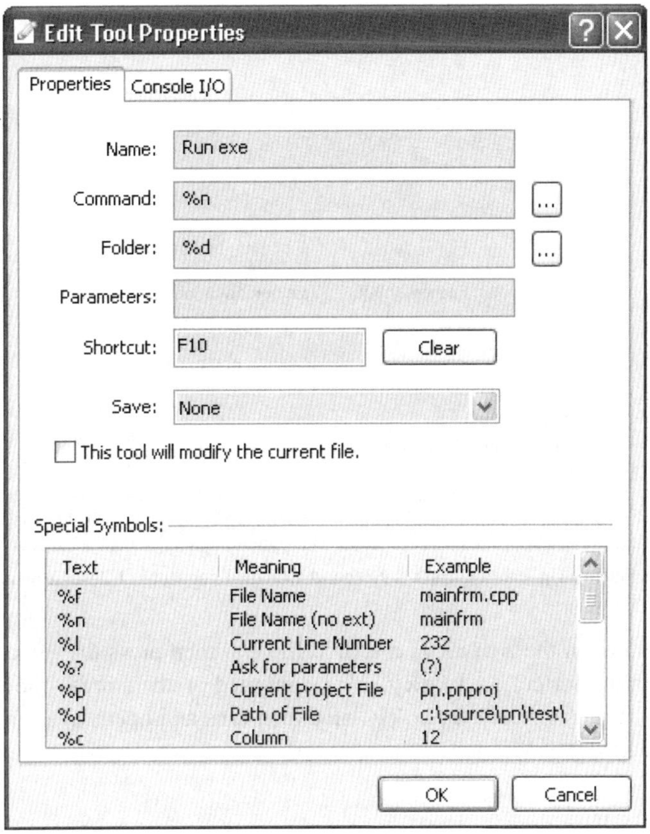

Figure 2.6: Setting Up Programmer's Notepad to Run Executable Files, Properties Tab

In order to compile or run a program, the source file of the program must be open in Programmer's Notepad. The source file for the program to run must also be the currently selected file (it is possible to have more than one source file open at the same time in Programmer's Notepad.) To run the program, first compile using your shortcut key combination, next run it by using your run shortcut keyboard combination or by clicking the **Tools** menu item and "**Run exe**" menu item that you created above.

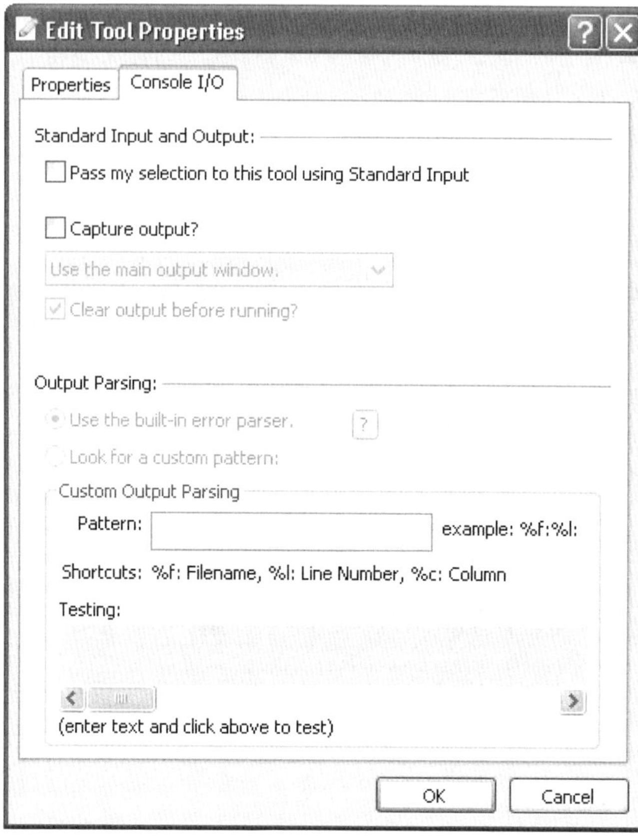

Figure 2.7: Setting Up Programmer's Notepad to Run Executable Files, Console I/O Tab

For example, if I open the **`sticky.c`** source file and then press the F9 key (my chosen *compile* keyboard shortcut), **`sticky.c`** is compiled. I then press the F10 key (my chosen *run* keyboard shortcut), **`sticky.exe`** then runs and opens in a window just as if I had double clicked it in Windows.

2.2 How the sticky.c Program Works

Now that we have finished setting up the tools, let's examine the **`sticky.c`** program. First note how the program runs inside **`main()`**. Each statement is executed (run) in order from top to bottom. The top **`printf()`** statement is run first, printing out the text "**`The console window will stay open.`**" to the screen. It has two newline characters at the end of the text. The first one moves the cursor onto the line below the text that is printed out, the second one moves the cursor to the line below that – resulting in a gap of one line between the first line of text and the next one.

The second **printf()** statement is run next, printing "**Press any key to close the window.**" to the screen at the current cursor position. There are no newline characters in this text, so when you run the program, the cursor will be flashing at the end of this line of text.

The third statement in the program contains a new function called **getch()**. What this function does is it gets a character from the user. A character is any keyboard character that you type. The function will wait for the user to type a character and this is why the program will not exit and disappear when you run it by double clicking it in Windows.

The only other thing new to you in this program is the line **#include <conio.h>**. For now all you need to know is that this line must be included at the top of your source file if you want to use the **getch()** function.

2.3 Input, Output and Variables

All programs work with input and output. The **hello** program worked with output only by displaying a message on the screen. The **sticky** program used output and input – output to display a message on the screen (the program output a message to the screen) and input in that it reacted to a keyboard key press input by the user. These programs are hardly useful in real life as they are just meant to illustrate how programming works while keeping the programs very simple.

Most programs will receive some input, do processing on the input and then produce an output that is useful. An embedded system may read the temperature (input), decide whether the temperature is too high or too low and then switch off or on a heating element (output) as in **figure 2.8**.

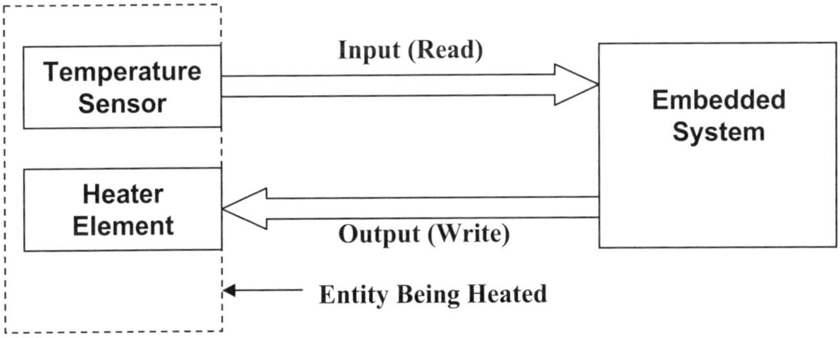

Figure 2.8: Example of Input and Output in an Embedded System

If we are to get an external measurement or value into the computer to do some processing on, we need to store the value in the computer's memory so that we can then use it in our computer program. The value that is read is stored as a variable.

A variable is a piece of memory set aside to store a variable value as opposed to a fixed or constant value. A temperature is an example of a variable value. When it is read by the computer it is not known by the computer and could be any value within a range. If I say that the computer reads an external temperature, what is happening is that the computer is running a program and the program tells the computer to read a value. The act of reading the value is the act of getting the value and storing it in memory as a variable.

To demonstrate input, output and variables on the PC in a useful way, we will get a number from the user of the program (typed in by the user on the keyboard), perform addition with the number and then output the result to the screen. Before we can do this, we need to be able to:

1. get a value that the user types in on the keyboard

2. store the value on the computer so that we can add a number to it

3. be able to display the result of the addition to the user on the screen.

Let's start by looking at how to print a number to the screen. We'll use the **printf()** function again. Create a new C source file in Programmer's Notepad and add the following lines of code:

```
#include <stdio.h>
#include <conio.h>

int main(void)
{
    printf("Printing a number: 19\n");
    printf("Printing a number: %d", 19);

    getch();
    return 0;
}
```

Save the program as **number.c**, compile and run it. From now on, you can run your programs by double clicking them in Windows so long as you use the **getch()** function to stop the console window from disappearing. You can also run them from within Programmer's Notepad if you set it up to do so. The PC sample programs in this book will all include this function. You can close the program by pressing any key on the keyboard.

The program shows that you can print a number as part of the text as with the first **printf()** function or as an external number that gets substituted into the text. In the second **printf()** function, the number 19 gets substituted in place of the **%d** format specifier. We will look at format specifiers in more detail in later programs. We will need the second method of printing a number that is variable (in this program the number is a constant and we are substituting a constant into the string – it is hardcoded into the program and can't be changed after the compile).

The first way of printing the number will not allow a stored value entered by a user of the program to be displayed on the screen because it is fixed as part of the text string (it is known as a string constant). We will substitute a stored variable value into the string just like the number 19 is substituted into the second text message.

The number program might not make much sense right now, as the two **printf()** statements produce exactly the same output although they do so in a different way.

2.3.1 Variables

As already explained, variables in C can store (or hold) a value that is input by the user. We can also store the result of a calculation in a variable. If you were to add two numbers together in a program, the result would need to be stored somewhere so that it can be displayed to the user – it is stored as a variable. Variables can store different kinds of data and each kind of data must be specified when declaring a variable as the next program will illustrate.

Create the following program called **var.c**:

```
#include <stdio.h>
#include <conio.h>

int main(void)
{
    int myvar;

    myvar = 8;
    printf("The value of variable myvar is: %d", myvar);

    getch();
    return 0;
}
```

In this program, an integer variable is defined: **int myvar;**. This tells the computer that it must make space in memory for a variable of type integer (**int**). The variable is given a name (**myvar**) so that we can identify it and use it in our program.

An integer is a whole number (not a fraction) e.g. 5, 1, 94, 0, -3, 783, -102. The name of the integer can be anything that you choose. It may not begin with a number (0 to 9), may not contain any spaces (use the underscore character _ instead if you choose) and may not contain any special characters like !, +, *, #, ? or dots as most of these have special meaning in C. You may not use any of the C language keywords as a variable name either as these words are reserved and have special meaning in C.

Or put another way, a variable name can consist of any alpha-numeric characters and the underscore character, but may not start with a numeric character or consist of a C keyword. It is also not advisable to start the variable name with the underscore character as this is used by the compiler for internal variable names.

Syntax highlighting, as seen in Programmer's Notepad, is helpful in identifying keywords in C. So if a word that you choose to use as a variable name changes colour after typing it in, you will know that it is a keyword and not to use it.

The variable is assigned a value of eight in the program: **`myvar = 8;`**. This means that **`myvar`**, now contains the value eight – it is a location in memory and it is storing the value eight. The name of the variable is only used in the C source code. When the program is compiled and run, the variable will be allocated a space in the computer's memory and will be referenced by its address in memory.

In the **`printf()`** statement, the value of the variable **`myvar`** i.e. 8 is substituted into the text string in place of the format specifier **`%d`** by putting the name **`myvar`** after the comma.

In the next program we will define a variable for storing the result of an addition calculation.

Create the following program called **`add.c`**:

```
#include <stdio.h>
#include <conio.h>

int main(void)
{
    int result;

    result = 3 + 4;
    printf("The result of 3 + 4 is: %d", result);

    getch();
    return 0;
}
```

Here we see the addition operator (**+**) being used to add two numbers together. The result of adding 3 and 4 is stored in an integer variable named **`result`**. The value that this variable is holding is then printed to the screen.

We now have the ability to store a value as a variable, perform addition and print the variable to the screen. We need one last thing before we can write our input/output program – the ability to get a number from the user of the program.

2.3.2 Getting Input from the User – the **scanf()** Function

The **scanf()** function is used to get input from the user. Like the **printf()** function, the **scanf()** function uses format specifiers to specify what type of data it is using. We have only seen one format specifier so far (**%d**) which is used with decimal integer numbers.

To see the **scanf()** function in action, create the following program called **input.c**:

```
#include <stdio.h>
#include <conio.h>

int main(void)
{
    int input;

    printf("Enter an integer number: ");
    scanf("%d", &input);
    printf("\n\nYou entered %d", input);
    getch();
    return 0;
}
```

When you run this program, it will ask you for an integer number. Type an integer number and press **Enter**. The program will then display the number that you typed.

The **scanf()** function waits for the user to enter a number and then stores the number in the variable **input**. After the number is stored in the variable, it can be printed to the screen as in the previous program.

scanf() requires that the address of the variable is passed to it and this is done with the address operator (**&**). Remember that when the variable is defined, it is allocated a space in memory. The memory location has an address and by using the ampersand symbol (**&**), we are supplying the **scanf()** function with the address of the variable. This is a requirement of the **scanf()** function. If you used the variable name without the ampersand, **scanf()** would then be getting the value that the variable holds instead of the address of the variable. If you do not fully understand this, do not worry, it will become clear later.

If you want to use the **scanf()** function in your programs, you need to have the line **#include <stdio.h>** at the top of your source file – the same as for the **printf()** function.

Although this program is called **input.c**, it actually uses input and output functions, but does nothing to change the input value. In the next program we will achieve our earlier objective of receiving input, processing it and then producing output.

This program is called `inout.c`:

```
#include <stdio.h>
#include <conio.h>

int main(void)
{
    int input, result;

    printf("Enter a number to add 5 to: ");
    scanf("%d", &input);
    result = input + 5;
    printf("%d + 5 = %d", input, result);
    getch();
    return 0;
}
```

Two integer variables are used in this program – `input` and `result`. They are both defined on the same line and separated by a comma. They could also be defined on two separate lines as follows:

```
int input;
int result;
```

The program asks the user to enter a number. The number obtained from the user by the `scanf()` function is stored in the `input` variable, `5` is then added to the value stored in `input` and the result of this calculation is stored in the variable `result`.

The values contained in both `input` and `result` are passed to the `printf()` function. Note that we have two format specifiers (`%d`). The variable `input` is written first in the `printf()` function, so it gets substituted into the first format specifier. The variable `result` is written second and gets substituted into the second format specifier as illustrated in **figure 2.9**.

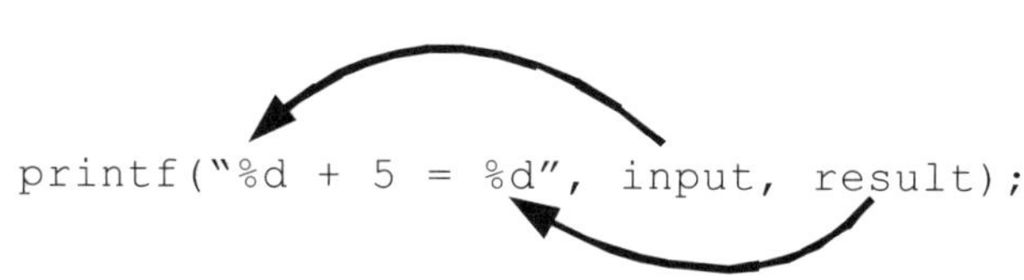

Figure 2.9: Substitution of Variable Values into Format Specifiers

2.4 Variable Types

So far we have used only integer variables, but there are other types of variables. The next program called **intdiv.c** will illustrate the necessity for a different type of variable.

```
#include <stdio.h>
#include <conio.h>

int main(void)
{
    int result;

    result = 3 / 4;
    printf("3 divided by 4 equals %d", result);
    getch();
    return 0;
}
```

This program displays the result of a division calculation using the division operator (**/**), but the result is not as expected. We would expect the result of **3 ÷ 4** to be **0.75**, but the result shown is **0**. This is because we are doing the division with integers and storing the result of the calculation in an integer variable.

Integers are whole numbers and in the computers memory can't hold fractions like the result of our calculation, so the fractional part of the result of the calculation is discarded. If we divided **7** by **4**, the result would be **1** instead of the expected **1.75**, again this is because the fraction is discarded.

To overcome this problem, we use a different variable type called a floating point variable.

2.4.1 Floating Point Variables

Floating point variables are able to store fractional numbers as the next program, **fltdiv.c**, shows.

```
#include <stdio.h>
#include <conio.h>

int main(void)
{
    float result;

    result = 3.0 / 4.0;
    printf("3.0 divided by 4.0 equals %f", result);
    getch();
    return 0;
}
```

The output from this program when it is run is now:

```
3.0 divided by 4.0 equals 0.750000
```

A variable of type **float** (floating point variable) named **result** is defined in the program. This variable is able to hold a floating point number. When we want to specify a constant floating point number in C it must contain a decimal point as in **3.0** and **4.0** in this program.

To print out the floating point result using **printf()**, the format specifier for floating point numbers is used - **%f**, as can be seen in the program.

2.4.2 Character Variables

Character variables are used to store individual characters as the next program **charvar.c** shows.

```
#include <stdio.h>
#include <conio.h>

int main(void)
{
    char ch;

    printf("Type a character and then press Enter: ");
    scanf("%c", &ch);
    printf("\nYou typed %c", ch);
    getch();
    return 0;
}
```

A character variable named **ch** is defined in order to store a character typed in by the user. The **scanf()** and **printf()** functions both use the **%c** format specifier in order to work with the character data type.

To initialize a character variable with a value in a C program, the character is written between single quotes (apostrophes). For example:

```
char ch;
ch = 't';
```

The first line defines a character variable with the name **ch**. The character **t** is then stored in the variable **ch**.

There is more to be said about variables, but we will use our knowledge gained so far to write some useful programs that do mathematical calculations.

2.5 Arithmetic Operators

Ohm's law can be used to calculate the current flowing through a resistor if the voltage across the resistor and resistance of the resistor are known. **Figure 2.10** shows the circuit used to demonstrate this. The resistor could represent a light bulb or relay coil for example. If you were to add an LED to your embedded system, you would need to be able to use Ohm's law to calculate the value of the current limiting resistor needed by the LED.

To calculate the current flowing through the resistor, simply divide the voltage across the resistor by the resistance of the resistor. The formula is represented mathematically by the following equation:

$$\mathbf{I = V \div R}$$

where

I is current in Amps
V is voltage in Volts
and **R** is resistance in Ohms.

E.g. if the voltage of the battery is **10V** and the resistance of the resistor is **3Ω** (3 ohms), then the current flowing in the circuit is **10 ÷ 3 = 3.333 Amps**.

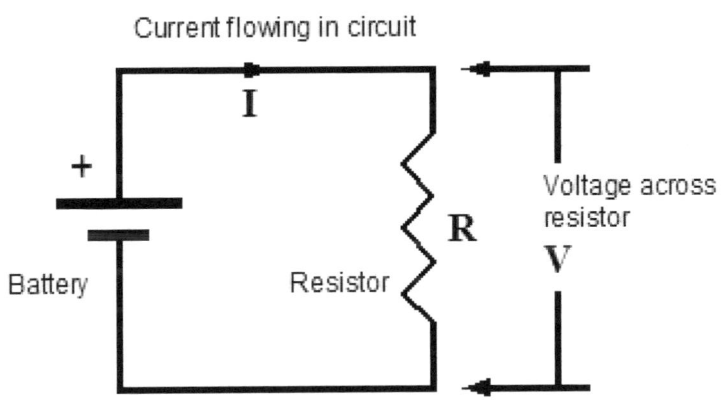

Figure 2.10: Circuit for Demonstrating Ohm's Law

The next program called **`ohm.c`** asks the user for the voltage and resistance and then calculates the current using the above equation.

```
#include <stdio.h>
#include <conio.h>

int main(void)
```

```
{
    float V, R, I;

    printf("Current calculation using Ohm's Law.\n");
    printf("\nEnter the voltage in Volts: ");
    scanf("%f", &V);
    printf("\nEnter the resistance in Ohms: ");
    scanf("%f", &R);
    I = V / R;
    printf("\n\nCurrent in Amps is %f A", I);
    getch();
    return 0;
}
```

A sample run of the program is shown below:

```
Current calculation using Ohm's Law.

Enter the voltage in Volts: 5

Enter the resistance in Ohms: 2

Current in Amps is 2.500000 A
```

The equals sign (=) is known as the assignment operator. The variable to the left of the assignment operator is assigned the value of whatever is on the right of it. In the above example, the variable **I** is assigned the result of the division of two variable values.
You have already seen the **+** (addition) and **/** (division) arithmetic operators. **Table 2.1** shows all of the arithmetic operators used in C.

Operator	Mathematical Symbol	Meaning	Example	Result
+	+	Addition	10 + 34	44
-	-	Subtraction	16 - 7	9
*	×	Multiplication	4 * 3	12
/	÷	Division	18.5 / 10.0	1.85
%	rem.	Remainder (modulo)	17 % 5	2

Table 2.1: Arithmetic Operators

The left column shows each operator as it is typed into the C source file using the keyboard.

Everything should be straight forward except for the modulo operator (**%**). The modulo operator gives the remainder left over from a division operation. In the example in the table, 5 will divide into 17 three times leaving a remainder of 2.

2.6 Field Width Specifiers

The output of the **fltdiv.c** program and the **ohm.c** program produce floating point numbers with 6 digits following the decimal point which is unnecessary. It is possible to change the number of digits displayed when using the **printf()** function. This is done by using a field width specifier as the next program, **fwidth.c**, shows.

```
#include <stdio.h>
#include <conio.h>

int main(void)
{
    printf("Float with 2 trailing digits: %.2f", 43.140000);
    printf("\nFloat with 3 trailing digits: %.3f", 43.140000);
    getch();
    return 0;
}
```

The output of this program produces:

```
Float with 2 trailing digits: 43.14
Float with 3 trailing digits: 43.140
```

As can be seen, by inserting **.2** between the % and the **f** in the format specifier, the number of digits displayed after the decimal point is reduced to two, and inserting **.3** reduces the number of digits to three.

2.7 Compiling and Linking

When using the C tools to create an executable file, there are actually two main steps involved. The first is called compiling and the second is called linking. We have been using the **gcc** program to compile and link in one step. The **gcc** program is actually a "front end" to the compiler and linker – it takes care of running the compiler and linker at the correct time.

It is important to know this because programming errors can come from the compile step and the linking step. Being aware of this will help you to find the source of the errors.

Figure 2.11 shows a diagram of the compile and link process. The diagram is using the first program that you wrote in chapter 1 to show what the tools do.

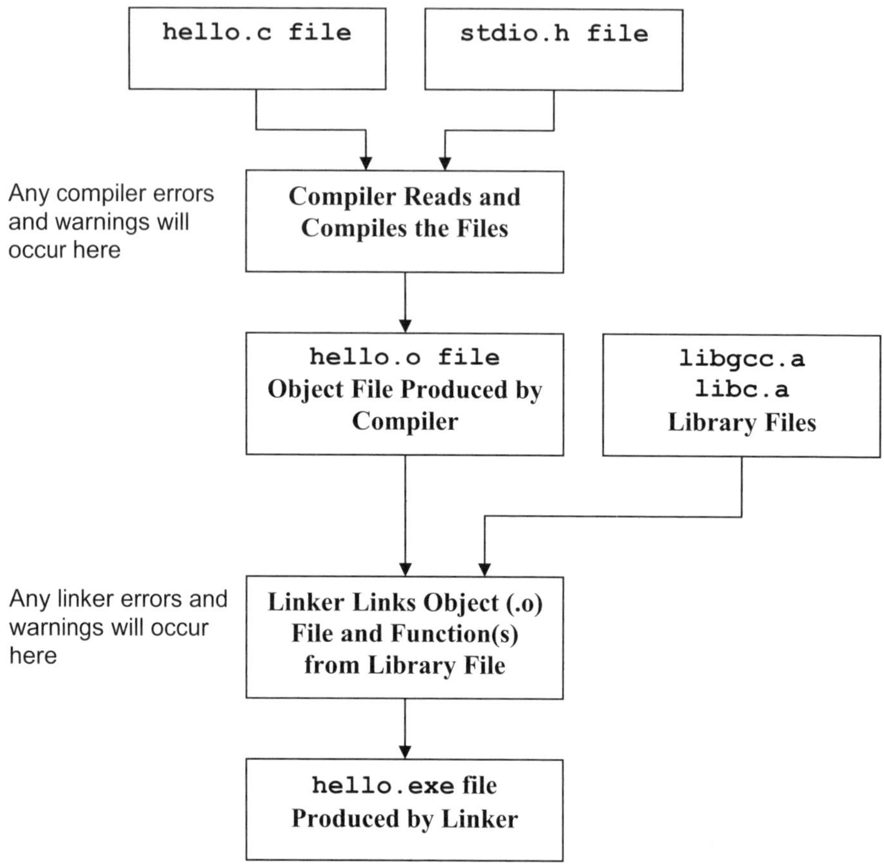

Figure 2.11: Compile and Link Process of the C Tools

The compiler compiles the **`hello.c`** file and the **`stdio.h`** file to produce an object file **`hello.o`**. The compiler knows that it must look for the **`stdio.h`** file because of the line at the top of the **`hello.c`** source file - **`#include <stdio.h>`**. This line is simply telling the tools to include this file in the compile step. The file **`stdio.h`** comes with the compiler and was installed when you installed the compiler. It can be found in the compiler directory: **`C:\DJGPP\include`**

We will not see the **`hello.o`** object file because the tools delete this file after linking is finished.

Remember that the **`printf()`** function that you used in the **`hello`** program exists externally as mentioned before. It actually exists in what is known as a library file. In the figure, two library files and the object file are linked together.

Linking is the process of adding the external functions used in the program to the executable file. The linker "resolves externals" – that means that it sees for example that the external function **printf()** was used in the program, it then looks for this function in object or library files and adds it when found. If it can't find the function, it will give an "unresolved externals" error and will not be able to produce the executable file.

The linker does not link the entire library file with the object file. It only links the functions that are used in the program.

The library files were installed with the compiler. **libgcc.a** can be found in **C:\DJGPP\lib\gcc\djgpp\4.23** and **libc.a** can be found in **C:\DJGPP\lib**. The tools are already set up to look in these directories for these files. **libc.a** contains the function **printf()** and **libgcc.a** contains some helper functions that the C program may need in order to run. (**libc.a** contains a lot of functions, **printf()** is just one of them. If any of the other functions were used in the program, they would be linked in and the rest left out)

The linker will produce an executable file as output. In the **hello.c** program this is the **hello.exe** file. This file is created by the linker and saved in a specific format that enables it to be run on a PC that is running Windows. If the same source file was compiled and linked with the Linux **gcc** tools on a Linux system, the linker would produce a file that is able to be run on the Linux system.

The ability of the C source file to be compiled and run on different systems is known as portability. Note that this only applies to the source file(s) and not the binary outputs of the compiler and linker (i.e. object and executable files). A compiler that is made for a particular system is needed each time.

2.7.1 Some Compile and Link Experiments

To better understand the compile and link process, it is possible to do these steps separately. Open up a DOS box using **djgppbox.bat** in the **C:\DJGPP** directory. In the DOS box, change to the directory that contains your **hello.c** file from chapter 1. Delete the **hello.exe** file if it exists: Enter **del hello.exe** at the command prompt to delete the file.

To compile **hello.c** without linking, type the following line at the command prompt:

```
gcc -c hello.c
```

Now enter **dir** or **explorer .** at the command prompt to see what is in the directory. You will see that this compile only step produced an object file called **hello.o**.

In the compile only step above, the **-c** that you typed is known as a switch. The switch is a message to the C tools to tell them to do something specific. To see a list of switches that you can use with **gcc**, enter **gcc --help** at the command prompt. You will see that next to the **-c** switch the following is displayed:

```
-c        Compile and assemble, but do not link
```

This explains why the **hello.c** file was compiled but not linked.

To perform the link step with the object file that has been produced, enter the following at the command prompt:

```
gcc hello.o -o hello.exe
```

This is very similar to the command that you typed to compile and link the C source file on its own, but here **gcc** sees that you are giving it an object file and it knows to run the linker and not to compile. This step will link the **hello.o** object file with any unresolved external functions in the library file.

If this does not make perfect sense to you right now, the main thing that you need to know is that the creation of an executable file from a C source file is a two part process – compiling and then linking. This is important to know when trying to figure out what error messages mean if you have an error in your program and it won't compile or link.

There are still some unanswered questions like: what exactly is the **stdio.h** file for? This can only be properly explained when we cover functions later in the book.

2.8 Errors and Warnings

Errors will prevent the compiler from compiling and producing an object file if the error is a compile error. If an error occurs when linking, the linker will not be able to produce an executable file.

Warnings will alert the programmer that something may be wrong in the program. The program will still compile and link, but may not work correctly. The compiler can be set up to treat warnings as errors and not produce any output if any warnings occur.

2.8.1 Compile Errors

Errors that occur during compiling are usually syntax errors such as a keyword being spelled wrong or a terminating semicolon that is left off.

Another example that will produce a compile error is if two variables are defined that have the same name. In C each variable defined must have a unique name.

The next program called **getchit.c** will teach you more about the **getch()** function that we have used before. It is a working program, but we will purposely put some errors into it to see what the compiler and linker will do.

```
#include <stdio.h>
#include <conio.h>

int main(void)
{
    char key;

    printf("Press a key: ");
    key = getch();
    printf("\nYou pressed the %c key\n", key);

    getch();
    return 0;
}
```

This program defines a character variable as you have seen before. After asking the user to press a key, the `getch()` function is called. It is possible to save the character that the user types on the keyboard using the `getch()` function. The character variable **`key`** is assigned the keyboard character that the user types by using the assignment operator (=) as in the line `key = getch();`.

In this line the function `getch()` "returns" the value of the key pressed by the user. The value is then put into the **`key`** variable.

Now remove the semicolon from the end of the variable definition line (`char key;`). Save the file and compile.

I get the following messages in the Output window in Programmer's Notepad:

```
> "C:\DJGPP\djenv.bat" getchit.c -o getchit.exe
getchit.c: In function 'main':
getchit.c:8: error: expected '=', ',', ';', 'asm' or '__attribute__' before 'printf'
getchit.c:9: error: 'key' undeclared (first use in this function)
getchit.c:9: error: (Each undeclared identifier is reported only once
getchit.c:9: error: for each function it appears in.)

> Process Exit Code: 1
> Time Taken: 00:01
```

This output shows that there is at least one error in the file. If you click any of the error messages in the **Output** window of Programmer's Notepad, the cursor in the source file will move to the line containing the error.

The first error message shows the source of the error. One of the missing items in the program is the semicolon as the message suggests. Because of this error, the compiler does not recognise that the variable **`key`** has been defined, hence the second error message.

The format of the error message identifies the name of the file that the error has occurred in – in this case **getchit.c**, and also the line number in which the error occurred – in this example lines 8 and 9. In the above example it appears that there are four errors, but looking at the line number shows that the last three messages all apply to the same error on line nine.

Fix the error by inserting the semicolon at the end of the variable definition again.

Now change the spelling of key in the line: **key = getch();**
Change it to: **ky = getch();**

After saving and compiling, the following error message is displayed:

```
> "C:\DJGPP\djenv.bat" getchit.c -o getchit.exe
getchit.c: In function 'main':
getchit.c:9: error: 'ky' undeclared (first use in this function)
getchit.c:9: error: (Each undeclared identifier is reported only once
getchit.c:9: error: for each function it appears in.)

> Process Exit Code: 1
> Time Taken: 00:00
```

The compiler does not recognise the identifier **ky**. It does not know what this is because it has not been defined to be a variable or anything else in the program, hence the error. The compile step can't be completed because the compiler does not know what **ky** is and so does not know what to do with it. The error has been identified as occurring on line 9.

Fix the error; we will look at link errors next.

2.9 Link Errors

To produce a link error, change the line:

```
printf("\nYou pressed the %c key\n", key);
```

to

```
prinf("\nYou pressed the %c key\n", key);
```

I.e. spell **printf** incorrectly.

This will produce a linker error because by spelling the name of the function incorrectly, we are referring to a function that does not exist. The compiler will actually warn you first that there is a function referred to that it does not know about. Because the linker looks for the function, it will produce the error message. The linker will try to find the function in one of the library files, but of course it will not find it.

When running the DJPGG **gcc** program from inside Programmer's Notepad to compile and link the faulty program, the following messages are produced:

```
> "C:\DJGPP\djenv.bat" getchit.c -o getchit.exe
getchit.c: In function 'main':
getchit.c:10: warning: implicit declaration of function 'prinf'
c:/djgpp/tmp/ccEOypnV.o:getchit.c:(.text+0x5e): undefined reference to
`_prinf'
collect2: ld returned 1 exit status

> Process Exit Code: 1
> Time Taken: 00:01
```

The warning message is from the compiler, but it still compiles correctly and produces an object file. The error message is from the linker and it fails to complete the linking process because of this error.

When the compiler is run automatically by **gcc**, it saves the output object file in a temporary directory with a filename that it has generated. This is what **c:/djgpp/tmp/ccEOypnV.o** is referring to in the error message.

So the linker message is saying that in the object file produced from the C source file **getchit.c**, there is a reference to **prinf** that is undefined – it can't find a function called **prinf** in the library files. The message also says that **ld** (the name of the linker program called by **gcc**) returned 1.

If a program returns 0, to the operating system or program that ran it, it means that the program succeeded in what it was asked to do. If the program returns anything else it means that the program failed. In this case the linker program returned 1, indicating that it failed to do the linking that it was asked to do.

If you can't tell if the error is from the compiler or linker, try compiling the file manually from the command prompt as we did earlier. If the compile does not produce any errors, then the error must be from the linker. You can try linking the object file manually as we did before to verify this.

2.10 Exercises

1. Write two program that use Ohm's law to calculate:
   a) The resistance of a resistor if the voltage across the resistor and the current flowing through the resistor is known.

   b) The voltage across the resistor if the current flowing through the resistor and the resistance of the resistor is known.

The program must ask the user to enter the two known quantities each time and then calculate the unknown quantity. The result must be displayed with two digits after the decimal point, use floating point variables throughout.

Use the following mathematical formulas to calculate the resistance and voltage. They have been derived by manipulating the formula that you saw earlier in the chapter:
To calculate resistance: **R = V ÷ I**
To calculate voltage: **V = I × R**

2. Which of the following are valid variable names?
   number8
   8th_number
   particle
   void
   big number
   small_number

   If you don't know the answer, try using them in a program and see what the compiler says.

2.10.1 Solutions

1. a)

```
#include <stdio.h>
#include <conio.h>
int main(void)
{
    float R, V, I;
    printf("Calculate resistor value:\n\n");
    printf("Enter voltage: ");
    scanf("%f", &V);
    printf("Enter current: ");
    scanf("%f", &I);
    R = V / I;
    printf("\n\nResistance is: %.2f Ohms", R);
    getch();
    return 0;
}
```

b)

```
#include <stdio.h>
#include <conio.h>
int main(void)
{
    float R, V, I;
    printf("Calculate voltage:\n\n");
    printf("Enter current: ");
    scanf("%f", &I);
    printf("Enter resistance: ");
    scanf("%f", &R);
```

```
    V = I * R;
    printf("\n\nVoltage is : %.2f volts", V);
    getch();
    return 0;
}
```

2. **Valid variable names:**
   `number8`
   `particle`
   `small_number`

   Invalid variable names:
   `void` – it is a C keyword
   `big number` – spaces are not allowed in variable names
   `8th_number` – variable names may not start with a number

2.11 Summary

- C program statements are run in order from top to bottom.

- Variables are used to store data in a program. The value that a variable is holding can be changed when the program is running.

- When defining a variable in C, the type of data that the variable will hold must be specified. (E.g. int, float, char)

- Constants are numbers that are "hard coded" into a program and can't be changed after compiling.

- A variable can be assigned the value of a constant, but the value of the variable can still be changed after this when a program is running. E.g.:
  ```
  int var;
  var = 25;
  ```
  `25` is a constant.
  `var` is a variable.
  `var` can be assigned a different value later in the program (re-used).

- Most programs will get input, perform a process on the input and produce an output.

- Arithmetic operators are used to perform basic mathematics on variables and constants.

- C source code is first compiled to produce an object file. The object file is then linked with functions from a library file in order to produce a working program.

- Errors and warnings from the C tools can occur during the compile and link stages.

This chapter has taken you through some of the basics of the C language. Almost every topic can be expanded on, but has been kept brief for now to increase your knowledge of C, but not overwhelm you with extra details that are unnecessary at this stage.

In the next chapter we will look at how a C program can be written to make decisions.

3 Comparative Operators and Decisions

What you will do in this chapter:

- Learn how to compare integer values
- Write programs that make decisions

3.1 Comparative Operators

Comparative operators are used to compare values. These values can be stored as constants or variables. For example is the value of variable X bigger than the value of variable Y? Comparative operators are also known as relational operators.

3.1.1 True and False

When an expression containing a comparative operator is evaluated, it will evaluate to either true or false. Logic systems and computers represent true by the value 1 and false by the value 0. The next example called **compare.c** will show what this means.

```
#include <stdio.h>
#include <conio.h>

int main(void)
{
    int temper = 20;

    printf("Is the temperature less than 23? %d\n", temper < 23);
    printf("Is the temperature more than 23? %d\n", temper > 23);

    printf("\nTemperature is now increasing...\n\n");
    temper = 50;

    printf("Is the temperature less than 23? %d\n", temper < 23);
    printf("Is the temperature more than 23? %d\n", temper > 23);

    getch();
    return 0;
}
```

In this example, a variable called **temper** is defined in order to store a temperature. The initial value assigned to the variable is 20, representing 20 degrees Celsius. As can be seen in the example, a variable can be defined and initialised with a value in one line; the examples in the previous chapter used two lines to do this. In the first **printf()** statement, the value stored in **temper** is compared to the constant value 23 using the mathematical "less than" sign: **temper < 23**. In other words we are saying "Is the value stored in the variable **temper** less than 23?"

When this expression is evaluated in C, it will evaluate to either true (1) or false (0). The result is then substituted in place of the format specifier. As can be seen when the program is run, the result of asking whether 20 (stored in **`temper`**) is less than 23 evaluates to 1 (true). Logically this is because 20 is less than 23. See **figure 3.1**.

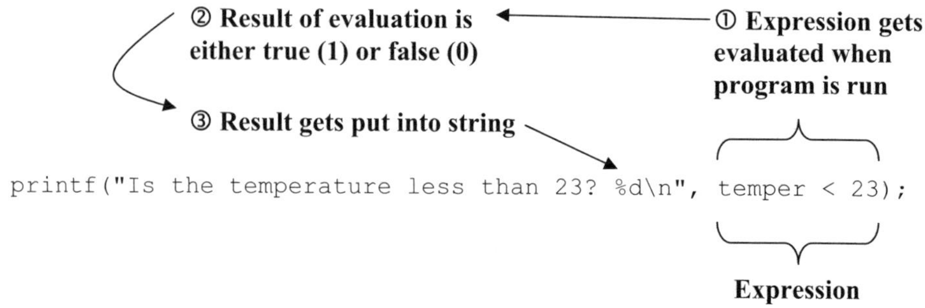

Figure 3.1: A Comparative Operator in Action

In the second **`printf()`** statement, we ask whether 20 is greater than (**>**) 23 and get a result of 0 (false).

When the value that **`temper`** is holding is changed to 50, and we ask the same questions, we get a result of false for our question "Is 50 less than 23?" and true for "Is 50 greater than 23?" as we would expect.

Table 3.1 shows the comparative operators available in C. You have already seen the less than and greater than operators in use.

Operator	Meaning	Example	Result (where 1 = true and 0 = false)
<	Less than	`A < B`	1 if A is less than B , else 0
>	Greater than	`A > B`	1 if A is greater than B , else 0
<=	Less than or equal to	`A <= B`	1 if A is less than or equal to B , else 0
>=	Greater than or equal to	`A >= B`	1 if A is greater than or equal to B , else 0
==	Equal to	`A == B`	1 if A is equal to B , else 0
!=	Not equal to	`A != B`	1 if A is not equal to B , else 0

Table 3.1: C Comparative Operators

3.2 Decisions

The comparative operators are used to make decisions in a C program. So far you have seen programs that execute (run) each statement in the program line for line from top to bottom. Decisions can be used to jump into or over a particular part of a program.

Getting back to our temperature controller example from chapter 2 (**figure 2.8**), a decision must be made to switch the element of the system on or off. The system needs to know if the temperature is too low and if it is, to switch the heater element on. When the temperature is too high, it must make a decision to switch the element off.

3.2.1 Using `if` to Make a Decision

The **`if`** keyword is used in a statement to make a decision as the next program **`whatif.c`** shows.

```
#include <stdio.h>
#include <conio.h>

int main(void)
{
    int temper;

    printf("Enter a temperature (0 to 100): ");
    scanf("%d", &temper);
    if (temper >= 30) {
        printf("\nSwitch the heater OFF");
    }

    getch();
    return 0;
}
```

This program asks the user to enter a temperature between 0 and 100. The **`if`** statement is used to determine if the temperature that was entered is greater than or equal to 30. If it is, a message is printed. If not the program skips the block of code between the braces under the **`if`** statement.

In an **`if`** statement, the expression between the parentheses must evaluate to true (1) in order for the block of code between the braces to be run. If the expression evaluates to false (0), the entire block is skipped.

When you run this program and enter a value that is below 30, it will appear that the program does nothing. This is because the **`getch()`** function is the next function that is run after the **`if`** statement is evaluated. Try modifying the program by adding a line above the **`getch()`** statement that prints a message such as: "End of program.".

3.2.2 Using `else` in Conjunction with `if`

The problem with using the **`whatif.c`** program for controlling a temperature is that it will not do anything to switch the heater on if the temperature drops below 30. The **`else`** statement is used to run a block of code when the **`if`** statement evaluates to false. The **`ifelse.c`** program will demonstrate how this is done.

```
#include <stdio.h>
#include <conio.h>

int main(void)
{
    int temper;

    printf("Enter a temperature (0 to 100): ");
    scanf("%d", &temper);
    if (temper >= 30) {
        printf("\nSwitch the heater OFF");
    }
    else {
        printf("\nSwitch the heater ON");
    }

    getch();
    return 0;
}
```

Now if the temperature is greater or equal to 30, a message is displayed to switch the heater off, otherwise the message to switch the heater on is displayed.

This program works the same as the previous one, except that should the `if` statement evaluate to false, the code in the `else` block will run, skipping the code in the `if` block. This is known as branching as the program will branch to the `else` statement. Should the `if` statement be true, the code in the block below the `if` statement will be run and the block of code under the `else` statement will be skipped over.

Figure 3.2 shows the program flow of the `ifelse.c` program.

In an embedded system the on and off messages can be replaced with the code that will actually switch a heater element on or off. The temperature of the quantity being heated would be obtained by reading a thermometer attached to the embedded system. The value obtained from the thermometer would be put into the `temper` variable.

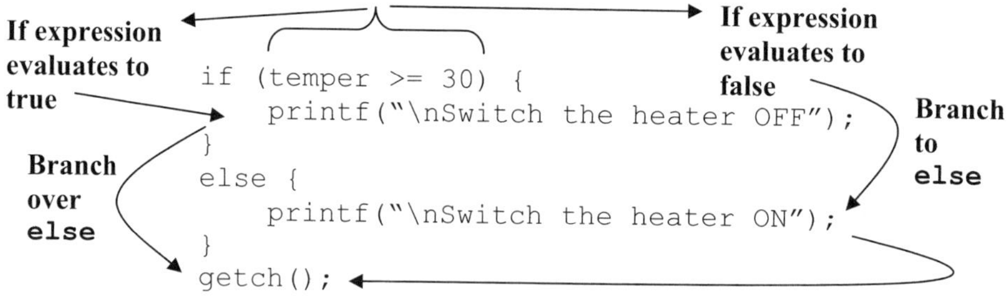

Figure 3.2: Program Flow in an `if` – `else` Construct

In the body (or block) of both the **if** and **else** statements, there can be multiple statements.
E.g.:

```
if (x < y) {
    statement1;
    statement2;
}
else {
    statement3;
    statement4;
}
```

When the **if** evaluates to true, statement 1 and statement 2 will run. When the **if** evaluates to false, statements 3 and 4 will run.

If the **if** or **else** blocks contain only one statement, then the enclosing braces of the body of the **if** or **else** is not required to have braces. The **if** – **else** construct used in **ifelse.c** could have been written like this:

```
if (temper >= 30)
    printf("\nSwitch the heater OFF");
else
    printf("\nSwitch the heater ON");
```

I find that a program is easier to read when the enclosing braces of a single line statement are in place, so I prefer to always put them in. This is a matter of personal preference and programming style.

3.2.3 The = Operator and the == Operator

It is important to know the difference between the assignment operator (=) and the comparative operator (==) (also known as the equality operator).

The assignment operator copies the value or result of an expression that is to the right of it into the variable to the left of it. e.g. **var = 3;**

The comparative operator compares the values to the left and right of it and if they are equal to each other, evaluates to true (1). If the values are not equal to each other, the comparison evaluates to false (0). e.g. **if (var == 3)**

In an **if** statement, any non-zero value will evaluate to true (1). For example:

```
if (24) {
    printf("This code will always run!");
}
```

When this **if** statement is evaluated, it will always evaluate to true because the number between the parentheses is non-zero.

It is also possible to use the assignment operator within the parentheses of the **if** statement as follows:

```
if (var = 3) {
    printf("This code will always run!");
}
```

The variable **var** is assigned the value of 3, **var** is therefore non-zero and will evaluate to true. This will result in the code within the **if** statement block always being run. This example code will compile and run and is perfectly legal in C, but may not be what the programmer intended.

This is a common error, where the programmer intends to compare the two values to see if they are equal, but ends up assigning a value to a variable instead.

The comparison of two constants by accidentally using the **=** assignment operator will cause the compiler to generate an error. If the value to the left of the **=** sign is a constant, the compiler will always generate an error.

Not possible to *assign* a constant to a constant

```
if (3 = 4) {
}
```

Legal to *compare* two constants (although the result of the comparison will always be the same)

```
if (3 == 4) {
}
```

Not possible to *assign* a variable to a constant

```
if (3 = var) {
}
```

Legal to *compare* a constant with a variable

```
if (3 == var) {
}
```

<u>Legal</u> to *assign* a constant to a variable

```
if (var = 3){
}
```

<u>Legal</u> to *compare* a variable with a constant

```
if (var == 3) {
}
```

The next program **fussy.c** will show how the equality operator works.

```
#include <stdio.h>
#include <conio.h>

int main(void)
{
    int temper;
```

```
    printf("Enter a temperature (0 to 100): ");
    scanf("%d", &temper);
    if (temper == 30) {
        printf("\nThe temperature is just right!");
    }
    else {
        printf("\nThe temperature is too hot or too cold.");
    }

    getch();
    return 0;
}
```

When run, the program asks the user for a temperature between 0 and 100. The **==** operator is used to check for an exact temperature of 30. Only if the user enters 30 will the block of code under the **if** statement be run. If any other value is entered, the code under the **else** block is run.

3.2.4 Using else if in Conjunction with if

If you want to write a program that checks for one of several possible values, you could use multiple **if** statements to do the checking. The program **multif.c** shows an example of this.

```
#include <stdio.h>
#include <conio.h>

int main(void)
{
    int input;

    printf("Enter 1 to start pump\n");
    printf("Enter 2 to start mixer\n");
    scanf("%d", &input);
    if (input == 1) {
        printf("Pump started.\n");
    }
    if (input == 2) {
        printf("Mixer started.\n");
    }
    getch();
    return 0;
}
```

In this program the user must enter either 1 or 2 to start a pump or mixer (represented by displaying a message on the PC). If the user enters 1, the "pump started" message will be displayed. The program will then check to see if the user entered 2, which is unnecessary because the value of **input** in this case is 1. If the variable **input** has the value of 1, it can't have the value of 2. If we wanted to add an **else** statement to take care of an incorrect number entered by the user, this would be difficult to check for. Thankfully it is possible to use the **else if** construct as shown in **figure 3.3**.

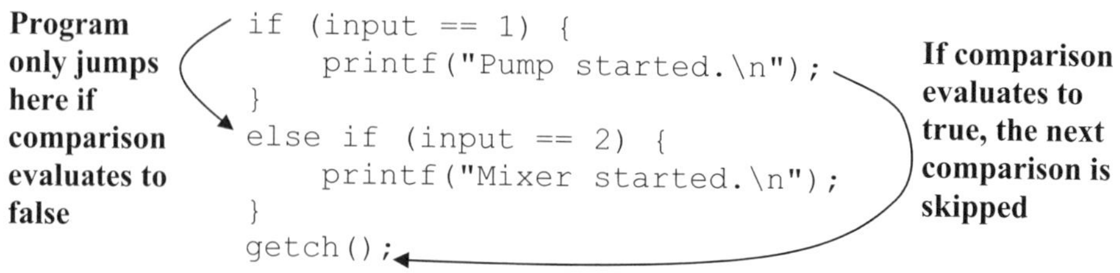

Figure 3.3: The `else if` Construct

In the **`else if`** construct, the comparison at the **`else if`** statement will be skipped, should the result of the first **`if`** statement be true. Should the first **`if`** statement be false, the comparison at the **`else if`** statement will be evaluated.

It is possible to have multiple **`else if`** statements after an initial **`if`** statement. A single **`else`** statement can be used at the end to run code if all of the previous evaluations were false. The **`pumpctrl.c`** program clarifies this.

```
#include <stdio.h>
#include <conio.h>

int main(void)
{
    int input;

    printf("Enter 1 to start pump\n");
    printf("Enter 2 to start mixer\n");
    printf("Enter 3 to stop pump\n");
    printf("Enter 4 to stop mixer\n");
    scanf("%d", &input);
    if (input == 1) {
        printf("Pump started.\n");
    }
    else if (input == 2) {
        printf("Mixer started.\n");
    }
    else if (input == 3) {
        printf("Pump stopped.\n");
    }
    else if (input == 4) {
        printf("Mixer stopped.\n");
    }
    else {
        printf("Invalid option\n");
    }
    getch();
    return 0;
}
```

In this program, the value of the variable **input** is checked for four different values. If none of the values are found, the code in the **else** block is run. If one of the values is found, the code in the block under the statement that evaluates to true will be run. After that, the program will continue to run at the **getch()** statement – all other statements in the **else if** construct are skipped.

3.3 Exercises

1. Write a program that compares two integer numbers of different values entered by the user and then tells the user which one is bigger. Hint: We have been comparing variable values with constants, but you can compare two variable values.

2. Modify the previous exercise to test if the first number entered by the user is greater than, less than or equal to the second number entered by the user. The result must be displayed to the user on the screen.

3.3.1 Solutions

1.

```
#include <stdio.h>
#include <conio.h>
int main(void)
{
    int num1, num2;
    printf("Enter a number: ");
    scanf("%d", &num1);
    printf("Enter a different number: ");
    scanf("%d", &num2);
    if (num1 > num2) {
        printf("\n%d is bigger than %d", num1, num2);
    }
    else {
        printf("\n%d is bigger than %d", num2, num1);
    }
    getch();
    return 0;
}
```

2.

```
#include <stdio.h>
#include <conio.h>
int main(void)
{
    int num1, num2;
    printf("Enter a number: ");
    scanf("%d", &num1);
    printf("Enter another number: ");
    scanf("%d", &num2);
    if (num1 > num2) {
        printf("\n%d is bigger than %d", num1, num2);
    }
    else if (num1 < num2) {
        printf("\n%d is smaller than %d", num1, num2);
```

```
    }
    else {
        printf("\n%d is equal to %d", num1, num2);
    }
    getch();
    return 0;
}
```

You could also check for **num1 == num2** by using an **else if** block in place of the **else** block above and leave the **else** block off altogether. This is, however, unnecessary as if a number is neither greater than nor less than another number, it must be equal to the other number. Alternatively you could firstly check if the two numbers are equal to each other, then if the first is greater than the second. The **else** block would then be run if the first number was less than the second. (Do any two of the checks in the **if** and **else if** statements and the **else** statement will take care of the third condition).

3.4 Summary

- When two values are compared by a comparative operator in C, they will evaluate to either true or false.

- In a computer, true is represented by 1 and false is represented by 0.

- Decisions can be made in C by using the **if** statement to evaluate a comparative expression.

- When the comparative expression inside an **if** statement evaluates to true, the block of code between braces under the **if** statement is run. Should the expression evaluate to false, the block of code under the **if** statement will be skipped over. The block of code is delimited by braces **{ }**.

- The **else** statement can be used to run a block of code when the **if** statement evaluates to false. Should the **if** statement evaluate to true, the block of code under the **else** statement will be skipped over (also known as branching or jumping – this also occurs when a single **if** statement is false; program flow branches over the **if** block).

- Any non-zero value between the parentheses of an **if** statement will evaluate to true.

- A common error in C is to accidentally use the = assignment operator instead of the == comparative operator when comparing two values to see if they are equal.

- **if** and **else** are both C keywords.

- The **`else if`** construct can be used to test for one of several conditions or values.

There are more ways to make decisions in C programs than what you have seen in this chapter. We will look at these in more detail when we are writing programs to run on an embedded system.

4 The `while` Loop

What you will do in this chapter:

- Learn how to use the **`while`** loop
- Improve the temperature controller example
- Learn about commenting a program and programming style

In the previous chapter we saw that a program can be made to branch when making a decision. Branching causes the regular flow of a program, from top to bottom through a list of statements, to change. Loops also change the flow of a program as we shall see.

4.1 The `while` Loop

The **`while`** loop will cause a block of program code to run repeatedly from top to bottom *while* a certain condition is true. Running the code repeatedly is called looping.

The **`water.c`** program will demonstrate the **`while`** loop:

```
#include <stdio.h>
#include <conio.h>

int main(void)
{
    int water_level = 0;

    printf("Pumping 10L of water...\n\n");

    while (water_level < 10) {
        water_level = water_level + 1;
        printf("Progress: %d\n", water_level);
    }

    printf("\nFinished.");
    getch();
    return 0;
}
```

The **`while`** loop evaluates a comparative expression to see if it is true or false in the same way as the **`if`** statement did in the previous chapter. The block of code between the braces under the **`while`** loop will run continually until the condition that it is evaluating becomes false.

When the loop is run for the first time, the expression **`water_level < 10`** is evaluated. Because **`water_level`** was initialised to the value of 0, the result of the evaluation is true (1). The code under the loop will now run because of this. If the evaluation had been false, the program flow would have ignored or "jumped over" the entire block of code inside the loop and carried on running any statements that occurred after this.

The first line of code that runs in the loop adds 1 to the value of the variable **`water_level`** and then stores this result back into the **`water_level`** variable (0 + 1 = 1). This result is printed to the screen.

Program flow has now reached the bottom of the loop, so the **`water_level < 10`** expression is evaluated again. The value of **`water_level`** is 1, and 1 is less than 10, so the evaluation result is again true and the first statement within the loop block is run again.

The running of this first statement again results in the **`water_level`** variable now containing the value 2 (1 + 1 = 2). As before, the result is printed to the screen and the loop expression is evaluated again.

This process will continue looping and when it does it increments **`water_level`** by one and prints the result to the screen each time.

When the value of **`water_level`** reaches 9, the loop expression will still evaluate to true. After the incrementing of this variable on the next pass through the loop, it will now be 10. The value of 10 is printed out to the screen. The loop expression gets evaluated again and evaluates to false because 10 is not less than 10. The code in the loop will now not be run again. Program flow continues at the **`printf()`** statement under the body of the **`while`** loop. **Figure 4.1** shows how the **`while`** loop works. Upon running the two statements in this loop, the expression will be evaluated again.

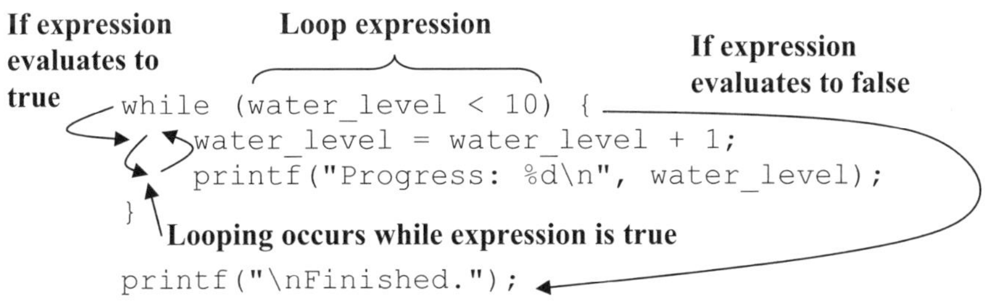

Figure 4.1: The `while` Loop

4.2 Using `if` Inside the `while` Loop

A **`while`** loop can be used to continually monitor a variable value as the following program, **`monit.c`** shows:

```
#include <stdio.h>
#include <conio.h>

int main(void)
{
    char key = 0;
    int num = 50;

    printf("Press u to increment number\n");
    printf("Press d to decrement number\n");
    printf("Press x to quit\n\n");

    while (key != 'x') {
        printf("Number is %d    \r", num);
        key = getch();
        if (key == 'u') {
            num = num + 1;
        }
        else if (key == 'd') {
            num = num - 1;
        }
    }

    return 0;
}
```

When this program is run, the user can press the 'u' key on the keyboard to add 1 to a number (increment the number by one) or the 'd' key to subtract 1 from a number (decrement the number by one). The value of the number is initially set to 50 and the new value of the number is displayed on the screen each time that it is changed. The 'x' key can be pressed to quit the program. Note that you won't be able to use the program if your Caps Lock key is on because the program is looking for lowercase characters only.

In this program, two variables are defined and initialised with values – one is used to store the character of the key pressed by the user on the keyboard, the other is used to store the count value used in the program. The three **`printf()`** statements at the top of the program simply give the user instructions on how to use the program.

The code in the **`while`** loop will keep running so long as the variable **`key`** does not contain the character 'x'. The character variable **`key`** has been initialised to the value 0 at the start of the program, so when the comparative expression of the **`while`** loop is evaluated, it is checked to see if **`key`** is not equal to (**`!=`**) 'x'. In this case **`key`** is not equal to 'x', it is equal to 0, so the answer to the question "Is **`key`** not equal to 'x'?" is true and the **`while`** loop is entered. Remember from chapter 2 that an individual character is written between single quotes in C (in the section "Character Variables").

Upon entering the loop, the value of the variable **num** is printed to the screen. This variable was initialised to 50, so 50 is printed to the screen.

The **getch()** function is run next and will stop the program while it waits for the user to press a key. When the user does press a key, the value of the key is stored in the variable **key** and the next line of code is run.

The **if** statement checks if the user pressed the 'u' key and if so, adds 1 to the variable **num**. If the key pressed is not 'u', the program flow jumps to the **else if** statement.

The **else if** statement checks if the value of the key pressed is 'd'. If it is, 1 is subtracted from the variable **num**. Regardless of whether one of the previous statements evaluated to true or false, program flow will continue at the top of the **while** loop where the loop expression is evaluated again.

If the 'x' key was pressed, then the variable **key** would hold the character 'x'. When the loop expression is evaluated again, it would evaluate to false and the loop would be exited. By exiting the loop, the program will run the **return** statement and then exit.

If none of the keys 'u', 'd' or 'x' were pressed, but some other key was pressed instead, the variable **key** would now contain that character value, but the program would just continue running without exiting the loop.

In the **printf()** statement inside the **while** loop, a new escape sequence is used, namely **\r**. This inserts a "carriage return" character into the end of the string. The "carriage return" causes the cursor to move back to the beginning of the line in which it occurs when the program is run. Run the program and notice that the cursor is positioned under the 'N' of the string "Number is 50". This is because of the "carriage return" character. It is used in this program to prevent the text from being printed across the screen and filling up the screen. Try removing it and see what happens.

4.3 The Guess My Number Game

This next program uses the **if** – **else** construct inside a while loop. The program is called **guess.c**:

```
#include <stdio.h>
#include <conio.h>

int main(void)
{
    int guess;

    printf("Try to guess my number:\n\n");
    printf("Enter a number between 1 and 100: ");
```

```
    scanf("%d", &guess);

    while (guess != 74) {
        if (guess > 74) {
            printf("Number is Lower.\n");
        }
        else {
            printf("Number is Higher.\n");
        }
        printf("Enter a number between 1 and 100: ");
        scanf("%d", &guess);
    }
    printf("You guessed the number!\n\n");
    printf("Press any key to quit.");
    getch();
    return 0;
}
```

The object of this program is for the user to guess a number. When run, the program prompts the user to enter a number between 1 and 100. The number is then compared to a constant value inside the program. If the user's number is higher or lower than the programs value, a message is displayed to tell the user if his number is higher or lower. If the number entered by the user matches the program's number, a message is displayed to say that the number guessed was correct.

To really make this program into a proper game, the program should generate a random number for the user to guess, rather than have a fixed value programmed into it. We will skip this step to keep the program simple.

An initial number is obtained from the user before evaluating the loop expression. If the user guessed the correct number the first time, i.e. 74, the loop expression would evaluate to false and the entire loop would be skipped over.

If the first number that the user entered was not correct, the **while** loop would be entered into. The **if** – **else** construct is used to tell the user if the number was too high or too low.

The user is then asked to enter a new number at the bottom of the loop. If the number is correct, the loop would be exited. If not the loop code will carry on running until the user enters the correct number.

4.4 Back to the Temperature Controller Example

In our temperature controller example from chapters 2 and 3, the temperature is read by the program and a message is displayed to either turn on or off an element. The problem with the program is that the temperature is only read once. We can now use a loop to continually monitor the temperature and display the on or off message as the temperature changes.

Our next program will allow the user to increase or decrease the temperature using the 'u' and 'd' keys on the keyboard. The program will keep monitoring the temperature and if it increases above a threshold value will display a message to switch the heater off. If below a threshold value the message to switch the heater on will be displayed.

The **montemp.c** program will show how this is done:

```
#include <stdio.h>
#include <conio.h>

int main(void)
{
    int temper = 30;
    char key = 0;

    printf("Press u to increase the temperature\n");
    printf("Press d to decrease the temperature\n");
    printf("Press x to quit the program\n");
    printf("Temperature is %d deg. C  \r", temper);

    while (key != 'x') {
        key = getch();
        if (key == 'u') {
            temper = temper + 1;
        }
        else if (key == 'd') {
            temper = temper - 1;
        }
        printf("\rTemperature is %d deg. C   ", temper);
        if (temper > 35) {
            printf("Switch element OFF.     ");
        }
        else if (temper < 30) {
            printf("Switch element ON.      ");
        }
        else {
            printf("Temperature is correct.");
        }
    }

    return 0;
}
```

There is nothing new to you in the **montemp.c** program, just more statements in the **while** loop. Because the computer is not interfaced to the thermometer and heater element, the user must play the role of the heating system that the computer is controlling. The user must control the heater element (by changing the temperature using the u and d keys) and must respond to the on and off messages as a heater controlled by an embedded system would.

The program is designed to keep a temperature between 30 and 35 degrees Celsius. If the user (heater) increases the temperature above 35, the message to switch the element off is displayed (the embedded systems switches the heater off). Of course it is up to the user to play the part of the heater and start decreasing the temperature when the heater is switched off by pressing the 'd' key a number of times to simulate the temperature decrease of the system.

When the temperature is decreased below 30, a message is displayed to switch the heater on. It is up to the user to now increase the temperature as the heater would and thus keep the temperature between the upper and lower limits (35°C and 30ºC).

When controlling the heater with an embedded system, the program would be written to read a thermometer continually and if the temperature was too high, switch the element off. If the temperature was too low, it would switch the element on. There would be no waiting in the loop as our example program does - waiting for a key press from the user.

4.5 Commenting Programs

Comments can be written into source code files by the programmer to explain what the program does. These comments are useful when the programmer wants to modify the program at a later date and may have forgotten exactly what the program code does. If another programmer is going to use or modify the program, he might not know what the author of the source code was intending to do. Comments help to explain what the program does. Note that comments should be used to explain what the program does, but not to explain how the programming language works.

Comments are written between a forward slash and asterisk (**/***) and an asterisk and forward slash (***/**) e.g.:

```
/* This is a comment */
```

Comments are included in the source code file and when compiled, the compiler ignores them. Comments can be written across multiple lines in a source code file e.g.:

```
/* This comment
occurs on
multiple lines */
```

Many C compilers allow the use of C++ comments. C++ comments can only be used on a single line and start with a double forward slash (**//**). There is nothing to terminate this style of comment.

The next program called **comment.c** will clarify how comments are used.

```
#include <stdio.h>
#include <conio.h>

/* C style comment */
/*
 Multi line C
 style comment
*/
```

```
// C++ style comment

// Let's see how to comment a program:

/*------------------------------------------
  comment.c
  This program converts kilometres to miles
  ------------------------------------------*/

int main(void)
{
    float kilos;    /* stores distance in kilometres */
    float miles;    /* stores distance in miles      */

    /* prompt the user for distance in km's */
    printf("\nEnter distance in kilometres: ");
    scanf("%f", &kilos);
    /* convert kilometres to miles */
    miles = kilos * 0.62137;
    /* show the user the result of the calculation */
    printf("\nDistance is %.3f in miles.\n", miles);

    getch();    /* wait for a keypress before closing window */
    return 0;
}
```

It is important to comment programs, particularly long ones. Commenting programs can save you hours of frustration when you come back to a program months after writing it and you can't remember what it does or how some part of the program works.

Comments should include such things as what units variables are stored in. In the above example, the comments clearly state what the variables are storing (distance) and in what units they are being stored (miles and kilometres).

Other information can also be added to a file in the form of comments such as who the author of the file is, the date it was written, when it was modified, who modified it, what hardware was the program written to run on, etc.

The example programs in this book will contain few comments to save space in the book and to help keep the code uncluttered. Some of the later more advanced programs will show good use of comments.

4.6 Programming Style

Programming style refers to how you write your programs in terms of where you put brackets and braces, how you indent your program, how many spaces are used for indenting and others.

As an example, the programs that you have seen in this book have been indented by four spaces. Indenting refers to the moving of statements to the right inside a block of code.

The following program (**style1.c**) will compile without any problems (included header files left out for clarity):

```
int main(void)
{
int count = 0;

printf("Counting to 15, skipping 10:\n");
while (count <= 15) {
/* if the number is 10, skip over it */
if (count != 10) {
printf("Count is: %d\n", count);
}
count = count + 1;
}
printf("Finished.\n");
getch();
return 0;
}
```

The above program becomes easier to read when it is indented (**style2.c**):

```
int main(void)
{
    int count = 0;

    printf("Counting to 15, skipping 10:\n");
    while (count <= 15) {
        /* if the number is 10, skip over it */
        if (count != 10) {
            printf("Count is: %d\n", count);
        }
        count = count + 1;
    }
    printf("Finished.\n");
    getch();
    return 0;
}
```

This program can even be written on a single line. See **style3.c** in the example programs from the accompanying zip file (appendix B). The file will compile and run just as the two previous examples did. (Scroll to the right in this version of the program to see the entire program.)

Other aspects of style refer to where you place the braces as the next examples show around an **if** statement.

As you have already seen:
```
if (count != 10) {
    printf("Count is: %d\n", count);
}
```

Some programmers prefer this style:
```
if (count != 10)
{
    printf("Count is: %d\n", count);
}
```

The next example is only legal if there is a single statement under the **`if`**. Some programmers prefer this and will then choose one of the above styles if there is to be more than one statement under the **`if`**.
```
if (count != 10)
    printf("Count is: %d\n", count);
```

This is also legal for a single statement:
```
if (count != 10) printf("Count is: %d\n", count);
```

Or this:
```
if (count != 10) {printf("Count is: %d\n", count);}
```

Some parts of coding style are just personal preferences, others are required to make a program clear, easy to read and understand. Bear this in mind when developing your own programming style.

4.7 Exercises

1. Write a program called **`subtract.c`** which asks the user for two numbers. The second number must be subtracted from the first and the result displayed. Use a loop to ask the user to either quit the program or do another calculation.

2. The example program **`pumpctrl.c`** from the previous chapter has a problem in that the program exits after a selection is made by the user. Modify this program to run inside a loop and continually ask the user for an option. Add an extra menu option that the user can use to exit the program.

4.7.1 Solutions

1.
```
#include <stdio.h>
#include <conio.h>

int main(void)
{
   int num1, num2;
   char key = 0;

   printf("Subtraction program\n");
```

```
    while (key != 'x') {
        printf("\nEnter a number: ");
        scanf("%d", &num1);
        printf("Enter a number to subtract: ");
        scanf("%d", &num2);
        printf("%d - %d = %d\n", num1, num2, num1 - num2);
        printf("\nPress any key to subtract again\n");
        printf("or x to quit\n");
        key = getch();
    }
    return 0;
}
```

The above solution to the exercise is one possible solution. You could have used floating point variables if you chose to and used a different key press to quit the program. Note that the subtraction can be done in the **`printf()`** statement. The program creates a variable internally to hold the result of the subtraction in this case. You could of course create a variable to hold the result of the calculation.

2.
```
#include <stdio.h>
int main(void)
{
    int input = 0;

    while (input != 5) {
        printf("Enter 1 to start pump\n");
        printf("Enter 2 to start mixer\n");
        printf("Enter 3 to stop pump\n");
        printf("Enter 4 to stop mixer\n");
        printf("Enter 5 to exit\n");
        scanf("%d", &input);
        if (input == 1) {
            printf("Pump started.\n");
        }
        else if (input == 2) {
            printf("Mixer started.\n");
        }
        else if (input == 3) {
            printf("Pump stopped.\n");
        }
        else if (input == 4) {
            printf("Mixer stopped.\n");
        }
        else if (input == 5) {
            printf("Exiting.");
        }
        else {
            printf("Invalid option\n");
        }
    }
    return 0;
}
```

Notice in the above solution that I chose not to use the **getch()** function. It is therefore not necessary to have **#include <conio.h>** at the top of the source file. The only problem with not using **getch()** is that the "Exiting" message would not be seen if running the program from Windows. Of course if you ran it from a command line interface window manually, you would be able to see the message.

4.8 Summary

- It is possible to alter the flow of a program by using a loop.
- The **while** loop will loop through the lines of code contained between its braces as long as the loop expression remains true.
- After the last line of code in the loop has been executed, the loop expression will be evaluated again.
- A loop can be used to continually monitor an external parameter (such as temperature) in an embedded system.
- Comments can be added to source code to explain what a program does as well as other information such as who the author of the program is, when it was created, what units a variable is in, etc.
- Programming style is a matter of personal choice - use it wisely.
- The main objective when developing your personal style of programming should be to make programs simple, clear, easy to read and understand.
- If you work for a software company, they may have a styles guideline document that will outline the preferred style that they would have you use when programming for them.

This chapter covered the **while** loop only. We will see that some other kinds of loops are available in C when we start programming our embedded system.

5 Functions

What you will do in this chapter:

- Learn how to write functions
- Use pre-processor directives
- Learn about header files
- Build a Project using the "make" utility
- See how functions relate to linking and library files

You have already used the **`printf()`**, **`scanf()`** and **`getch()`** functions. You have also written a function each time that you have written a C program – the function **`main()`**.

The **`printf()`** function was supplied with the compiler. The actual C code that makes up this function is rather large and complex as the function has a lot of text formatting capabilities. If you had to write this code each time you wanted to print text to the screen, programs would be very large. Hence you can see the advantage of writing functions. The code for the function is written once and the function is used many times over.

Another advantage of using functions is that your program can be split over multiple files. If a large program is stored in a single file, it would become very difficult to read. By breaking the program up into functions it is possible to split the program into two or more files.

When splitting up a program into files, similar functions can be grouped together. You may decide to save all of the functions to do with serial communications into a file called **`serial.c`** and the functions to do with timers to **`timer.c`**.

Code can also be easily reused when it is divided up in this way. For a particular piece of hardware, or algorithm you only need to write the functions once and then include the file that contains the functions in each program that you wish to write that uses them. For example we used **`printf()`** in all of our programs without ever writing the **`printf()`** function. If **`printf()`** had been a function that you had written, you would only have to create it once and then use it whenever you wanted to.

5.1 Your Second Function

This will be the second function that you are going to write as you have already written the function **main()** each time that you have written a C program.

The program **countfunc.c** will show you how to write a function.

```
#include <stdio.h>
#include <conio.h>

void Count(void);                       /* function prototype */

int main(void)
{
    printf("Counting to 5\n");
    Count();                            /* call the function */
    printf("Counting to 5 again\n");
    Count();                            /* call the function */
    getch();
    return 0;
}

/* the function */
void Count(void)
{
    int count = 1;                      /* count from 1 */

    while (count <= 5) {                /* count to 5 */
        printf("%d\n", count);          /* display the count */
        count = count + 1;              /* increment the count */
    }
}
```

In this example, a function called **Count()** is created and then called twice in the main program (**figure 5.1**). The function simply prints out a count of 1 to 5. Near the top of the source file is the "function prototype" terminated with a semicolon i.e. **void Count(void);**. The function prototype tells the compiler that this function will be used in the program, similar to defining a variable that will be used in a program. The **main()** function is a special function that does not require a function prototype.

The actual function is written below the closing brace of the **main()** function. The body of this function is contained between opening and closing braces.

The use of the function in this example saves the programmer from having to write the same code twice to print out the count sequence. The function is written once and can be called any number of times in the program.

The first use of the C keyword **void** in front of the function name tells the compiler that the function does not return a value. The second use of the keyword **void** between the parentheses of the function tells the compiler that the function takes no arguments. We will see what this means in the examples that follow.

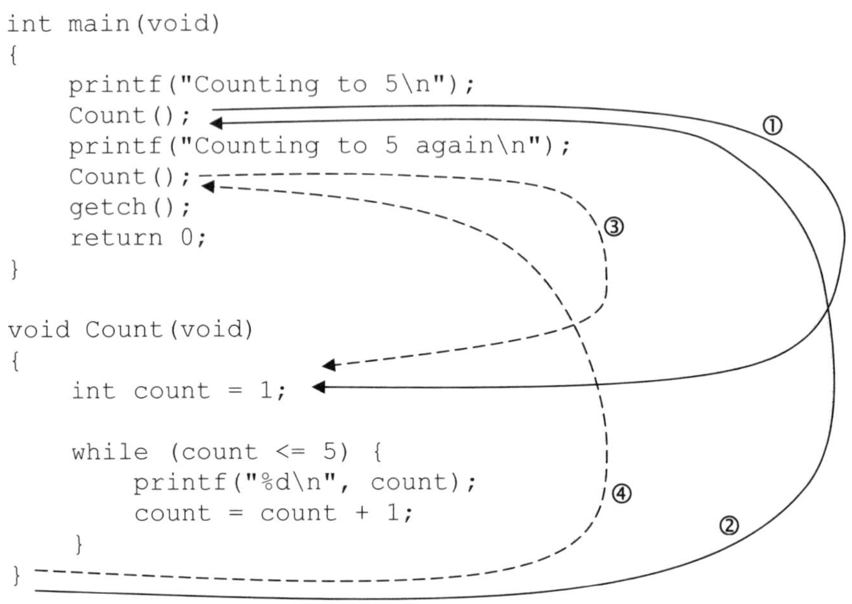

① First call to function
② Return from function after executing it
③ Second call to function
④ Return from function after executing it

Figure 5.1: How a Function is Called

5.2 Passing Data to a Function

It is possible to send data to a function by using arguments. The function is then able to operate on the data as the next program **passcount.c** shows.

```
#include <stdio.h>
#include <conio.h>

void Count(int num);

int main(void)
{
    printf("Counting to 2\n");
    Count(2);                          /* count to 2 */
    printf("Counting to 10\n");
    Count(10);                         /* count to 10 */
```

```
    getch();
    return 0;
}

void Count(int num)
{
    int count = 1;                      /* count from 1 */

    while ( count <= num) {             /* count to num */
        printf("%d\n", count);          /* display the count */
        count = count + 1;              /* increment the count */
    }
}
```

This program is very similar to the previous one except that the value that the function is to count to is passed to it as an argument. When the **`Count()`** function is called for the first time, a value of 2 is passed to it (the function takes an argument). The value of 2 is stored in the variable **`num`** in this function. When this variable is used in the function, the count is printed to the screen and the **`while`** loop exited when a count greater than 2 is reached.

When **`Count()`** is called the second time, 10 is passed to it and the **`while`** loop inside this function will now count to the new value of **`num`** which is now 10. So the variable between parentheses of the function takes on the value that is passed to it – in this example the variable **`num`**.

5.3 Passing More Than One Value to a Function

In the previous example, a single value was passed to the function (the function took a single argument). It is possible to pass more than one value to a function. The **`twoargs.c`** program contains a function that takes two arguments:

```
#include <stdio.h>
#include <conio.h>

void Count(int num, int skip);

int main(void)
{
    printf("Counting to 5, skipping 3\n");
    Count(5, 3);                    /* count to 5, skip 3 */
    printf("Counting to 12, skipping 10\n");
    Count(12, 10);                  /* count to 12, skip 10 */
    getch();
    return 0;
}

void Count(int num, int skip)
{
    int count = 1;                  /* count from 1 */
```

```
    while ( count <= num) {          /* count to num */
        if (count != skip) {
            printf("%d\n", count);   /* display the count */
        }
        count = count + 1;           /* increment the count */
    }
}
```

In this program, two values are passed to the **Count()** function. The first value is passed to it in the variable **num** and the second in the variable **skip**. The variable **num** is used by the function as the value to count up to and the variable **skip** contains the number that the function will skip over when printing out the count sequence. When calling the function it is important to put the two numbers that are to be passed to the function in the correct order. If the first call to the function had been **Count(3, 5);** then the **num** variable would have been assigned the value 3 and the **skip** variable the value 5. In this case the function would just count to 3 as there is not a value of 5 in this sequence to skip over.

As you have seen in the previous two examples, values are passed to a function by defining variables between the functions parentheses. If more than one value is to be passed to the function, the variables are separated by commas. A function that is to receive four values (take four arguments), for example, will be written like this:

```
void MyFunc(int val1, int val2, float val3, char val4)
{
    // the integers val1 and val2 can be used by the function
    // the floating point variable val3 can be used by the function
    // the character variable val4 can be used by the function
}
```

As this example shows, the type of data to be passed to the function is specified when declaring the function arguments. Each of these different data types are passed to the function and can be used by the function. Each new variable to be passed to the function between parentheses is separated by a comma.

When calling this function from within a program, values passed to it must be in the correct order (two ints first, one float and then one char):

```
MyFunc(100, 20, 43.45, 'z');
```

5.4 Passing a Variable to a Function

In the preceding examples, we have passed a constant value to the functions. It is also possible to pass a variable value to a function as in **vararg.c**.

```
#include <stdio.h>
#include <conio.h>
```

```
void Message(char ch);

int main(void)
{
    char key = 0;

    printf("Press a key\n");
    printf("Press x to exit\n");
    while (key != 'x') {            /* exit if x key pressed */
        key = getch();              /* get key press from user */
        Message(key);               /* display message */
    }

    getch();
    return 0;
}

void Message(char ch)
{
    if (ch == 'x')
        printf("Bye!\n");
    else
        printf("You pressed %c\n", ch);
}
```

A character variable **key** is passed to the **Message()** function in this example. The value of **key** is copied into the functions variable **ch** when the function is called.

5.5 Getting a Value Back from a Function

Functions can return a value as we have already seen the **getch()** function do. The type of value being returned must be specified (e.g. int, float, char) and can be returned by the function by using the **return** keyword. **retval.c** will demonstrate:

```
#include <stdio.h>
#include <conio.h>

int prompt(void);

int main(void)
{
    int user_num;

    user_num = prompt();
    printf("You entered %d\n", user_num);

    getch();
    return 0;
}

int prompt(void)
{
    int num;
```

```
    printf("Please enter an integer number:\n");
    scanf("%d", &num);
    return num;
}
```

When the **prompt()** function is called, it asks the user of the program to enter an integer number. The number is then stored in the local variable **num** (local to the function, it can't be "seen" by the **main()** function) . The number is then returned to the **main()** function by using the **return** keyword, where it is assigned to the variable **user_num** using the assignment (=) operator. The return type of the function is specified by placing the variable type keyword in front of the function name, in this case, **int**.

5.6 Passing Values to a Function and Receiving a Value Back

The next example program **addfunc.c** will receive two values passed to it and return a value.

```
#include <stdio.h>
#include <conio.h>

float AddNum(float num1, float num2);

int main(void)
{
    float add_result;

    printf("Adding two numbers:\n");
    add_result = AddNum(10.34, 100.56);
    printf("Result is %.2f", add_result);

    getch();
    return 0;
}

float AddNum(float num1, float num2)
{
    float result;

    result = num1 + num2;
    return result;
}
```

A function called **AddNum()** is used to add two numbers together and return the result of the addition. Floating point numbers are used throughout. The return type of the function is specified as floating point and the two numbers passed to the function are also specified as floating point. In this example you could, of course, have more easily just added the numbers together in **main()** without creating a function.

It is possible to return more than one value from a function, but you will have to wait until we see how pointers work in a later chapter.

5.7 Flashing LED Simulation Program

The **flashled.c** program simulates a flashing LED by displaying an on or off message instead of switching an LED on or off. In an embedded system the flow of the program will be exactly the same; the only difference being the code that displays the messages will be replaced by the code that switches the LED on and off. A function is called to use up some time while the LED is either on or off. If this function was not present, the LED would flash so fast that it would look like it was permanently on (it might look a bit dimmer) or on the PC the messages will look like they have merged. To exit the program, the user presses any keyboard key.

```
#include <stdio.h>
#include <conio.h>

void Delay(void);

int main(void)
{
    while (!kbhit()) {          /* press any key to exit */
        printf("ON \r");        /* switch LED on */
        Delay();                /* leave LED on for a while */
        printf("OFF\r");        /* switch LED off */
        Delay();                /* leave LED off for a while */
    }
    return 0;
}

void Delay(void)
{
    int count = 100000000;

    /* waste time */
    while (count) {
        count = count - 1;
    }
}
```

The **Delay()** function works by counting to a very big value which uses up processor (CPU) time. This leaves the message displayed on the screen for the duration of the delay before it is changed to a different message. You may have to experiment a bit with the value that is assigned to **count** in the **Delay()** function. This is because your PC may be running at a different clock speed to mine. The PC may also be running other programs in the background, slowing it down. I found that the messages were rather erratic – not displaying the on or off messages for the same length of time. This is because Windows is not a "real-time" operating system. It will run the program, but is not guaranteed to run it at exactly the right time as it may be servicing other programs that are running on the system. I found that making the console window full screen helped. To make the console window full screen, run the program, hold down the **`Alt`** key and then press the **`Enter`** key. To switch back to windowed mode, hold down the **`Alt`** key and press the **`Enter`** key again (**`Alt`** + **`Enter`**). You may have to experiment by changing the delay value again when using full screen mode.

A new function called **`kbhit()`** is used in this program. This function will return true if a key is pressed on the keyboard and false if it is not. Unlike the **`getch()`** function, it does not wait for a key press as this would hold the program up at the top of the loop. Each time it is called, it just returns 1 (true, a key was pressed) or 0 (false, a key was not pressed).

The **not** operator (`!`) is used in front of the **`kbhit()`** function to change (invert) the value returned by this function. If the function returns true, the **not** operator changes the result to false, if the function returns false, the **not** operator changes it to true. This is required in order to stay in the **`while`** loop. We want to stay in the **`while`** loop while a key has not been pressed. Calling the **`kbhit()`** function without the **not** operator will return false because a key will not have been pressed when the program starts. The false return value will prevent the code inside the **`while`** loop from being run.

You have already seen the **not equal to** operator (`!=`) looking for the opposite of the equality (==) operator. The **not** operator is similar in that it inverts a true or false value.

Another thing new in this program is the calling of a function within the **`while`** loops parentheses in **`main()`**. This is perfectly legal in C. The result of the function call is evaluated to be true or false as with any expression in the loops parentheses. In this case the not operator modifies the returned result from the **`kbhit()`** function as part of the evaluation.

5.8 Pre-processor Directives

You have already seen the **`#include`** pre-processor directive in each program that you have written. The **`#include`** pre-processor directive is used to include a header file (a text file with a **`.h`** file extension). These files contain the function prototypes for functions such as **`printf()`** and **`getch()`**. As you have learned, each function must have a function prototype. By including the header files, you include the function prototypes for the **standard library functions** that are part of the C language standard. Using the **`#include`** pre-processor directive to include a header file is like copying the contents of the header file and pasting them into the top of your C source file.

The **C standard library** containing the **standard library functions** has a number of header files associated with it. The library file is linked with your object file during the linking process. The header files contain the function prototypes of the library functions that you use from the library file (**`.lib`**).

For example, the function prototypes for **printf()** and **scanf()** are in the **stdio.h** header file. The function prototypes for **getch()** and **kbhit()** are in the **conio.h** file. **stdio** stands for "standard I/O" (I/O = input/output). **printf()** and **scanf()** are standard I/O functions for a PC. **conio** stands for "console I/O" and deals with input and output from the console window (DOS box). These are not the only functions in the library or function prototypes in the header files, there are a number of other functions available.

While on the subject of names, the functions that you have been using from the C standard library are shortened names that describe the function:

printf() = print formatted (there are lots of formatting options that can be used with this function like printing floats or ints, etc.).

scanf() = scan formatted

getch() = get character

kbhit() = keyboard hit

Back to pre-processor directives – there are additional pre-processor directives other than **#include**. Each pre-processor directive starts with the hash (**#**) symbol.

The **#define** pre-processor directive can be used to name a constant value used in a C program. For example the line:

```
#define MY_NUM 100
```

placed at the top of a C source file will enable you to use the text **MY_NUM** everywhere that you wish to use the value 100.

The name defined by the **#define** directive is usually written in upper-case letters by most C programmers, although it is possible to use any combination of upper and lower-case letters. The rules that apply to variable names when defining variables also apply to names defined with **#define**, e.g. the name must start with an alphabet or underscore character, etc.

In **defdel.c**, the previous program has been modified as follows:

```
#include <stdio.h>
#include <conio.h>

#define DEL_CNT      100000000

void Delay(int count);

int main(void)
```

```
{
    while (!kbhit()) {
        printf("ON \r");        /* switch LED on */
        Delay(DEL_CNT);         /* leave LED on for a while */
        printf("OFF\r");        /* switch LED off */
        Delay(DEL_CNT);         /* leave LED off for a while */
    }
    return 0;
}

void Delay(int count)
{
    /* waste time */
    while (count) {
        count = count - 1;
    }
}
```

The **Delay()** function has been modified to take the delay count value as an argument. The **#define** directive has been used to define **DEL_CNT** to the value of 100000000. 100000000 will now be substituted into the program everywhere that **DEL_CNT** is used before the file is compiled. This is done by the pre-processor.

The pre-processor is either a separate program that is run before the compiler or a process that is part of the compiler but is run before compiling. The pre-processor will substitute the header files into the C source file when it finds the **#include** directive or substitute the constant value in place of the name used with the **#define** directive.

So the pre-processor will change this (extracted from the above program):

```
int main(void)
{
    while (!kbhit()) {
        printf("ON \r");        /* switch LED on */
        Delay(DEL_CNT);         /* leave LED on for a while */
        printf("OFF\r");        /* switch LED off */
        Delay(DEL_CNT);         /* leave LED off for a while */
    }
    return 0;
}
```

So that the compiler sees this:

```
int main(void)
{
    while (!kbhit()) {
        printf("ON \r");        /* switch LED on */
        Delay(100000000);       /* leave LED on for a while */
        printf("OFF\r");        /* switch LED off */
        Delay(100000000);       /* leave LED off for a while */
    }
    return 0;
}
```

The **#define** directive as used in the above program enables the programmer to change the value passed to the **Delay()** function at the top of the source file, no matter where or how many times it is used in the program. This saves the programmer from having to search for each occurrence of the constant value in the program in order to change it. It can be changed once at the top of the file.

Another use of the **#define** directive is to define a mathematical constant. For example the constant PI (π) used in mathematics has the value of 3.14159. A program using this value becomes a lot more self explanatory when using the letters PI in place of the number 3.14159.

The mathematical formula for calculating the area of a circle is:

$$A = \pi r^2$$

where:
A = area
π = 3.14159
r = radius of the circle (distance from the centre of the circle to the edge)

r^2 means r × r

This formula would be written in C as:

```
area = 3.14159 * radius * radius
```

circle.c shows how to define PI to make the program more readable.

```
#include <stdio.h>
#include <conio.h>

#define PI  3.14159

float CircArea(float radius);

int main(void)
{
    float radius;

    printf("Enter radius of circle: ");
    scanf("%f", &radius);
    printf("Area of circle is %.3f", CircArea(radius));

    getch();
    return 0;
}

float CircArea(float radius)
```

```
{
    return PI * radius * radius;
}
```

PI is defined using the **#define** directive as previously described and used in the **CircArea()** function to calculate the area of a circle when given the radius of the circle.

Previously you always defined a variable when doing a calculation. If you look at the **CircArea()** function, no variable has been defined in which to store the result of the area calculation before it is returned by this function. This will still work in C. A place to store the result before returning it is still needed, but when a program is written like this one, a temporary variable is created in the program and is not seen by the programmer.

Another new thing in this program is calling the **CircArea()** function directly in the **printf()** function. This is also perfectly legal in C and a temporary variable will again be created for you and passed to the **printf()** function when you write a program in this way.

Notice also that a floating point variable **radius** is defined in **main()**. A floating point variable named **radius** is also defined as an argument of the function **CircArea()**. In C you can not have two variables of the same name in a program. Because the two variables are defined in two separate functions, it is not a problem if they have the same name. The code running in **main()** and using the variable **radius** does not know about the variable with the same name in the **CircArea()** function. This is called the scope of the variable. The variables defined in **main()** can only be "seen" and used in **main()**. The variables defined in any other function can only be "seen" and used in that function. When the **CircArea()** function is called in the above program, the variable **radius** in **main()** is copied into the **CircArea()** functions variable **radius** as would occur if the variables had different names.

5.9 Functions Calling Functions

You can write functions that call other functions. This is useful when breaking up a very big program into smaller parts to make it easier to read and manage. The **main()** function has already been used to call library functions as well as other functions that you have written. The program that follows will clarify. You don't have to type this program in, it does nothing.

```
void a(void);      // prototype for function a
void b(void);      // prototype for function b

int main(void)
{
    a();           // main() can call function a()
    b();           // main() can call function b()
```

```
}

void a(void)
{
    b();          // function a() can call function b()
}

void b(void)
{
}
```

5.10 Using Multiple Source Files

It is possible to split a C program into multiple source code files by using functions. Functions used in a program do not have to exist in the same source file as **main()**. One reason for splitting up a program into more than one file is to make a large program modular by grouping similar functions together with each group in its own file. A large program becomes more maintainable this way than putting all of the source code in a single file. Another reason is to easily reuse source code by copying the files that contain the desired functions into another project.

Our compiling and linking will have to change a bit when using more than one source file. We will have to compile each source file individually. The object files produced by the compiling will all have to be linked together in the linking step. The **splitprog** program will demonstrate this. This program consists of two files: **splitprog.c** containing the function **main()** and **splitfunc.c** containing a function used by **main()**.

```
/* splitprog.c */
#include <stdio.h>
#include <conio.h>

/* prototype for function in splitfunc.c */
int FindBiggest(int num1, int num2);

int main(void)
{
    int first, second;

    printf("Program determines which number is bigger.\n");
    printf("Enter an integer number: ");
    scanf("%d", &first);
    printf("Enter a 2nd integer number: ");
    scanf("%d", &second);

    printf("%d is bigger", FindBiggest(first, second));

    getch();
    return 0;
}
```

The second file contains:

```
/* splitfunc.c */
/* function returns the bigger of two numbers */
int FindBiggest(int num1, int num2)
{
    if (num1 > num2) {
        return num1;
    }
    else {
        return num2;
    }
}
```

You will need to build this project manually using **djgppbox.bat** to start a console window. In the console window enter:

```
gcc splitprog.c splitfunc.c -o splitprog.exe
```

The above line compiles the two C source files and then links them. This could also be done in two separate compile and one link step, but the above line does it all for you. The separate compile and link commands look like this:

```
gcc -c splitprog.c

gcc -c splitfunc.c

gcc splitprog.o splitfunc.o -o splitfunc.exe
```

This will produce the two object files from compiling first and then link these object files to produce the executable file. Library functions will automatically be linked in.

5.11 Header Files

In the previous example, you could put the function prototype of the **FindBiggest()** function into a header file. The header file would then be included in **main()** by using the **#include** directive. The function prototype would then not have to be written at the top of the file containing **main()**. A header file called **splitfunc.h** would be created – create a C source file and save it with a **.h** file extension.

splitfunc.h would contain:

```
int FindBiggest(int num1, int num2);
```

The first few lines of **splitprog.c** would now contain:

```
// splitprog.c
#include <stdio.h>
```

```
#include <conio.h>
#include "splitfunc.h"

int main(void)
...
...
```

Note that when including your own header file, the header file name is placed between inverted commas (""). When using standard library header files, the header file name is placed between angle brackets (<>). These angle brackets are the same characters used as less than and greater than signs in C code.

Using a header file to contain a single function prototype as above is not very useful. When a C source file contains many functions, it makes much more sense to use a header file for the function prototypes. For a source file containing say ten functions, you will create the header file and then include it in every source file that uses one or more of the functions.

5.12 The make Program and Make File

When C projects get bigger and consist of many source code files, a utility program called **make** (**make.exe**) is often used to coordinate the build process of the project. A text file containing build instructions for the project is created and the **make** utility reads the instructions from this "make file" and builds the project. You should have installed the **make** utility when installing the DJGPP compiler. To test that the **make** utility is present on your PC, start a console window with **djgppbox.bat** and enter the following:

```
make --version
```

A version number and copyright message will be displayed if **make** was installed.

The first embedded program that you write in this book will require the use of a make file and that is the reason for covering make files in this chapter. The next example will show the use of a make file. The project consists of two C source files (**mkmain.c** and **mkgraph.c**), a header file (**mkgraph.h**) and a make file (**makefile**). When the **make** utility is run, it opens the make file which instructs it to run the compiler to compile the C source files and then call the linker to do the linking to produce the **graphprog.exe** executable.

To run **make**, you will need to start a console window with **djgppbox.bat** and then change to the directory in which the source files and make file exist. My files are in the **CH5** sub-directory of my **cprogs** directory in which **djgppbox.bat** starts the console window. After typing in the contents of the four project files and saving them (or copying them from the zipped sources), you need to enter the following in the console window:

```
cd ch5
```

to change to the directory in which the source files are located. Change the directory name in the line above to reflect your directory name.

Then enter:

```
make
```

This will run the make utility program which will automatically look for a file called **makefile**. If you had chosen to name your make file **mkgraph** for example you would need to enter:

```
make mkgraph
```

to get **make** to open this make file.

The source code listing is below.

First source file containing **main()**:

```
// mkmain.c
#include <stdio.h>
#include <conio.h>
#include "mkgraph.h"

int main(void)
{
    GraphHeader();

    printf("\n\nOil:    ");
    PrintGraph(10);
    printf("\n\nFlour: ");
    PrintGraph(18);
    printf("\n\nSugar: ");
    PrintGraph(4);
    printf("\n\n");

    getch();
    return 0;
}
```

Second source file containing two functions used in **main()**:

```
// mkgraph.c
#include <stdio.h>

void GraphHeader(void)
{
    printf("\n----------------\n");
    printf("| Tank levels: |\n");
    printf("----------------\n");
}
```

```
void PrintGraph(int length)
{
    if (length < 0) {   // can't have a negative graph
        return;
    }
    if (length > 20) {  // maximum graph length = 20
        length = 20;
    }

    while (length) {    // draw the graph
        printf("%c", 176);
        length = length - 1;
    }
}
```

A header file containing the function prototypes of the functions in **`mkgraph.c`**:

```
// mkgraph.h

void GraphHeader(void);
void PrintGraph(int length);
```

The makefile:

```
graphprog.exe: mkmain.o mkgraph.o
	gcc mkmain.o mkgraph.o -o graphprog.exe

mkmain.o: mkmain.c
	gcc -c mkmain.c

mkgraph.o: mkgraph.c
	gcc -c mkgraph.c
```

NB: use **`Tab`** character each time

Note: When entering the text into the make file, it is important to use the **`Tab`** character to indent the lines rather than inserting a number of spaces when the **`Tab`** key is pressed. If you set up Programmer's Notepad to use spaces instead of **`Tab`** characters, when editing the make file you need to first click the **Tools** menu item and then **Use Tabs**. If you don't do this then your make file won't work. You will get a message similar to this:

```
makefile:2: *** missing separator.  Stop.
```

When I inserted four spaces at the first indent, I got the above message when typing **`make`** at the command prompt.

The **`graphprog.exe`** program displays bar graphs representing the quantity of various ingredients in a biscuit making factory. Functions to do with the graphs are grouped together in their own file called **`mkgraph.c`**. These functions are called in **`main()`** which resides in the **`mkmain.c`** file.

In the **`PrintGraph()`** function, the character used to print the bar graph comes from the ASCII table and is represented by the number 176. We will learn more about the ASCII table in a later chapter. This character is printed to the screen using **`printf()`**

and the character format specifier (`%c`). It is printed **`length`** number of times with the value of **`length`** being the value that is passed to **`PrintGraph()`** when it is called. The **`while`** loop will be run as long as the value of **`length`** is a non-zero number. This variable is decremented (one is subtracted from it) at each pass through the loop. When the number reaches zero, the loop expression evaluates to false and the loop is exited.

There are two **`if`** statements above the **`while`** loop that keep the value of **`length`** within a range. The first one causes the function to be exited by running the **`return`** statement if a negative number is passed to the function. This has been included in the program because a negative amount of an ingredient is meaningless in this case. The second **`if`** statement prevents the length of the graph from being more than 20. If the user of the function passed a number to it that was greater than 20, the function would make it equal to 20. This is to keep the bar graph from going off the end of the screen or keeping it at a required maximum length.

5.12.1 How the Make File Works

The make file consists of "dependency lines" containing a colon (`:`) and command lines each beginning with the **`Tab`** character. The dependency lines contain the target to be made to the left of the colon. The object file names and executable names are found here as they are the targets that are to be made. The lines beginning with the **`Tab`** character contain the instructions for making the target. To the right of each colon are the targets prerequisites for building it.

When run, the **`make`** utility will look at the targets prerequisites and see if the date of these files are newer than the targets date. If they are, the target will be re-built. If not, the dependency files have not been modified since the last build and therefore the target is up to date and does not have to be re-built. In this way the make utility will only build files for which source code has been modified. This saves time when building large projects that have many source files by only re-building those that have changed.

If you have already run **`make`** on the example project and then you modify one of the C source files, only the modified file will be re-compiled when running **`make`** again. The project will be linked again because the object file produced from compiling the modified C source file will have a newer date than **`graphproj.exe`**.

5.13 How Functions Relate to Linking and Library Files

After seeing how a program can be broken up into functions, you should now be able to understand the big picture of how the C tools work. The library files that get linked with your programs were written in C and then compiled. The object files that were produced by the compiling are all grouped together in the library files. When the C tools are installed on your system, these library files are supplied as well as header files that contain the function prototypes of the functions that are in the library files.

When you want to use a function from one of these library files, you include the header file that contains the function prototype in your source file and the linker then links the function into your program when it is built.

5.14 Exercises

1. Write a program to simulate two flashing LEDs. Instead of printing text on the screen, use the ░ and █ characters for LED off and LED on. The two LEDs must flash alternately i.e. when the first is on, the second must be off and vice versa.

   The number to use for the ░ character is 176. The █ character is 219. Use a function to delay the on and off times of the simulated LEDs.

2. Write a function that prints out a bar graph with the length of the bar graph being passed to the function as an integer. A calling program must receive the integer value from a user's key presses on the keyboard and pass it to the bar graph function. Choose a key that will increment the integer (add 1 to it) and a different key that will decrement the integer (subtract 1 from it) e.g. 'u' and 'd' for up and down. The calling program must keep the value passed to the bar graph function between 0 and 30. A different key press must be used so that the user can exit the program.

   This program is trickier to write than it sounds. When the bar graph is drawn, it must always be drawn from the left side of the screen. When the bar has been incremented and then decremented by the user, the old bar characters must be erased by printing spaces where they are no longer needed. See how far you can get and then take a peek at the solution if you can't get it to work. Then try it again yourself. Use the **`#define`** directive to define a name to use for the upper limit of the bar graph i.e. 30.

3. Modify the previous exercise to:
   a. Change the upper limit of the bar graph length to 35 by changing the value of the name defined by the **`#define`** directive.
   b. Call a function that makes the PC speaker beep if the user tries to increment the value passed to the bar graph function to a number greater than 35.
      To make the PC speaker beep, print the character with the value of 7. This character will not print, but make the PC sound a beep. I get a Windows default sound when the program is run in windowed mode and a different sound when run full-screen on my PC.

5.14.1 Solutions

1.
```
// altflash.c
#include <stdio.h>
#include <conio.h>

void Delay(void);
```

```
int main(void)
{

    while (!kbhit()) {
        printf("%c %c\r", 176, 219);    // LED 1 off, 2 on
        Delay();
        printf("%c %c\r", 219, 176);    // LED 1 on, 2 off
        Delay();
    }

    return 0;
}

void Delay(void)
{
    int count = 100000000;

    /* waste time */
    while (count) {
        count = count - 1;
    }
}
```

2.
```
// barctrl.c
#include <stdio.h>
#include <conio.h>

#define MAX_BAR_LEN 30

void BarGraph(int length);

int main(void)
{
    char key_press = 0;
    int graph_val = 0;

    printf("u to inc., d to dec., x to exit\n\n");

    while (key_press != 'x') {              // exit if x pressed
        BarGraph(graph_val);                // draw the graph
        key_press = getch();                // wait for key press
        if (key_press == 'u') {
            graph_val = graph_val + 1;  // bar length increases
        }
        else if (key_press == 'd') {
            graph_val = graph_val - 1;  // bar length decreases
        }
        if (graph_val > MAX_BAR_LEN) {
            graph_val = MAX_BAR_LEN;    // limit bar length
        }
        else if (graph_val < 0) {
            graph_val = 0;              // bar can't go negative
        }
    }
```

```
    return 0;
}

void BarGraph(int length)
{
    int max_length = MAX_BAR_LEN;

    printf("\r");                       // start bar on left
    while (max_length) {
        if (length) {
            printf("%c", 176);          // draw bar character
            length = length - 1;
        }
        else {
            printf(" ");                // space erases prev. bar
        }
        max_length = max_length - 1;
    }
}
```

3. In the solution above, change the value of **`MAX_BAR_LEN`** to 35 i.e.:

```
#define MAX_BAR_LEN 35
```

Add the **`Beep()`** function to the program (don't forget the function prototype):

```
void Beep(void)
{
    printf("%c", 7);
}
```

And then use it in the program:

```
if (graph_val > MAX_BAR_LEN) {
    graph_val = MAX_BAR_LEN;        // limit bar length
    Beep();
}
else if (graph_val < 0) {
    graph_val = 0;                  // bar can't go negative
}
```

5.15 Summary

- Functions can be used to break a large program into smaller units.
- Functions can reduce the size of a program by reusing code that is called often.
- Every function must have a function prototype, except for **`main()`**.
- Data can be passed to a function and received back from a function.
- The type of data sent to or received from a function must be specified.

- Pre-processor directives are read by and applied by the pre-processor before compilation.

- The **`#include`** pre-processor directive is used to include a header file in a C source file. The header file can contain function prototypes and other information.

- The **`#define`** pre-processor directive can be used to label a constant value. The constant value will be substituted in place of the label throughout the source file by the pre-processor.

- A C program can consist of more than one source file. This is made possible by splitting the program into functions.

- A multiple source file program can be compiled and linked on the command line, but a better way to build the project is to use a make file.

- The **`make`** program will run a make file that will check to see which source files of a project have been modified by comparing the dates of the object files with the dates of the corresponding source files. The make program will only build the target files who's corresponding source files have been modified.

- Functions can be supplied to a programmer in a library file with corresponding header files. The programmer does not necessarily have to have access to the source files of the functions when supplied in this way. This is the case with the C standard library supplied with the C compiler and containing functions such as **`printf()`** and **`getch()`**.

The main objective in this chapter was to teach you how to write your own functions. The section on using make files to build a project was an introduction into how larger projects are built. When starting to program embedded systems in chapter 8, you will immediately need to start using a make file to build your project. This is necessary because when programming an embedded system in C, an assembly language source file is needed to set up the microcontroller before running the C program. The make file is used to assemble the assembly file and compile the C file and then link the two together.

6 Number Systems

What you will do in this chapter:

- Learn why we use binary numbers
- See how the binary numbering system works
- Learn to read and use hexadecimal numbers
- Use hexadecimal numbers in C programs
- The ASCII Alphanumeric code

Everything in a computer or microcontroller is stored as a binary number, whether it is text, a program, a photograph or video. A computer only works with binary numbers. It is therefore important to know how number systems work and this chapter will explain what you need to know about numbers as an embedded C programmer.

6.1 Binary Basics

Here is an example of a binary number:

`11001000`

This number is said to be an 8 bit number as it consists of 8 digits or bits. Each binary digit is either a zero (0) or a one (1). The digit to the very right of the number is known as the least significant bit (LSB) or least significant digit (LSD). The digit to the very left of the number is known as the most significant bit (MSB) or most significant digit (MSD). This is always the case, no matter how many bits the number consists of.

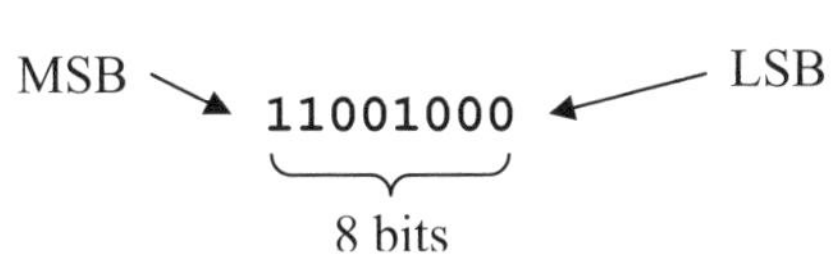

The word "bit" comes from the words **b**inary dig**it**.

When memory is referred to in relation to computers, it is usually specified in bytes e.g. 16 kilo bytes. A **byte** consists of 8 bits. A **nibble** is half a byte or 4 bits. The length of a **word** is dependent on the system architecture to which it applies and may be 16 bits or 32 bits for example. A word with a length of 32 bits is referred to as a 32 bit word.

1 bit = a single binary digit and can have a value of 0 or 1
1 nibble = a sequence of 4 bits
1 byte = a sequence of 8 bits
1 word = a sequence of 32 bits on an ARM system or 4 bytes

When a binary number is written in text, it is usually terminated with the letter 'B' to distinguish it from a decimal number or a number from a different numbering system. Here are some examples:

1 bit **1B**
1 nibble **1010B**
1 byte **00101111B**
1 16 bit word **1000100100001101B**

Some texts may use a lower case b at the end of a binary number e.g. **1001b**. Others may use a small 2 e.g. $\mathbf{11100101}_2$ where the 2 represents the two digits (0 and 1) of the binary system.

6.2 The Need for Binary Numbers

Numbers represented on a computer consist of voltage levels. Only two voltage levels are used: zero volts (0V) and a positive voltage e.g. five volts (5V). Binary numbers consist of only two digits zero (0) and one (1). A binary digit 0 is stored as a voltage level of 0 volts and a binary digit 1 is stored as a positive voltage, 5V for example.

If you could open up a memory chip that is attached to a computer system and could use a voltmeter (or multi-meter set to voltage) to measure each binary digit in the chip, you could write out what was stored in the chip. Every time you measured 5V you would write down a 1 and every time that you measured 0V you would write down a 0.

So a series of measurements as follows:
5V, 0V, 0V, 5V, 0V, 0V, 0V, 5V
would represent the binary number:
10010001

The actual voltage used to represent a binary digit 1 is dependent on the system that it is used in. In the past, digital systems always used 5V to represent the binary digit 1. Modern microprocessors usually run on lower voltages like 1.8V for example. On these systems a binary 1 would be represented by 1.8V. The memory interfaced to one of these systems would then have to consist of 1.8V memory chips.

6.3 Numbering Systems

6.3.1 A Quick Look at Decimal Numbers

To help understand how binary numbers work, we will have a quick review of the decimal numbering system.

The decimal number system should be familiar to you. Decimal numbers consist of ten digits 0 to 9 (1 to 9 = 9 digits, 0 to 9 = 10 digits). The decimal system is therefore said to have a base of ten. The use of only ten digits does not prevent us from counting higher than nine. When counting to a value larger than 9, we use two or more digits. The position of each digit in the number tells us what size the digit is. This is known as a weighted numbering system. For example the number 44 contains the same two digits (symbols), but their position in the number determines the value of each digit.

The weight of the each digit is a power of 10 increasing from right to left, with the first digit on the right being 1. For the example of the number 44, the first digit on the right has a weight of 1 and the second digit from the right has a weight of 10:

$4 \times 10 +$
4×1
$= 44$ (4 10s and 4 1s)

The "power of ten" is a mathematical term and is written mathematically as:

$10^0 = 1$	(ten to the power of zero equals one)
$10^1 = 10$	(ten to the power of one equals ten)
$10^2 = 100$	(ten to the power of two equals one hundred)
$10^3 = 1000$	(ten to the power of three equals one thousand)

In mathematics any number raised to the power of zero equals one, that is why $10^0 = 1$. Any number raised to the power of one is equal to itself: $10^1 = 10$. The number that ten is raised to in the above examples is the number of times that ten is multiplied by itself:

$10^0 = 1$ (any number raised to the power of 0 equals 1)
$10^1 = 10$ (any number raised to the power of one equals itself)
$10^2 = 10 \times 10 = 100$
$10^3 = 10 \times 10 \times 10 = 1000$

and so on.

The number 932 consists of 9 100s (10^2), 3 10s (10^1) and 2 1s (10^0):

$9 \times 100 +$
$3 \times 10 +$
2×1
$= 932$

The number 8157 can be viewed as follows:

Digit position from right	3	2	1	0
Weight	10^3	10^2	10^1	10^0
=	10 × 10 × 10	10 × 10	10	1
=	1000	100	10	1
Number	**8**	**1**	**5**	**7**
=	**8 × 1000**	**1 × 100**	**5 × 10**	**7 × 1**

Decimal numbers are written without anything appended to them to show what numbering system they are from e.g. 100 and 95 are decimal numbers whereas 100B represents a binary number. Decimal number are sometimes written with their base number included e.g. 191_{10} (one hundred and ninety one base ten) where 10 represents the ten digits that make up the decimal number system.

6.3.2 Binary Numbers

The binary number system uses only two digits as already explained (base two). It is also a weighted number system. When counting in decimal starting from 0, we move through the ten digits of the decimal system. After 9 is reached, the place holder to the left of the 9 is increased from 0 to 1 and the number becomes 10. Counting continues at 11 and when 19 is reached, the left most digit is incremented again so that the number becomes 20. This continues until 99 is reached and then the place holder to the left of this number is increased to 1 making the number 100.

Binary works in the same way, but there are only two digits to count through until the place holder to the left of the number must be increased by 1. We start with a count of 0 and count to 1. We are now out of digits so need to increment the place holder to the left of the number by 1. So the binary number becomes 10. The number to the right is now incremented and the number becomes 11. This would be equivalent of running out of digits when 99 is reached in decimal. In decimal the count changes to 100. In binary the same happens. The number now becomes 100, but the actual value is not read as "one hundred" when using binary it is read as "one zero zero" which has a count value of 4. After 100, the count continues to 101, 110 and 111. This may seem strange to you as it is a new concept, it is not difficult to understand but will require some getting used to.

You can use the Windows Calculator program to experiment with binary numbers. To start Windows Calculator click **start → All Programs → Accessories → Calculator** or in the Windows Run dialog box (**start → Run...**) type **calc** and hit **Enter**.

The first thing that you will need to do before using the calculator to work with binary numbers is to change it to scientific mode. In the calculator program click the **View** menu and then **Scientific**. Now click the **Bin** radio button to put the calculator into binary mode (or click the **View** menu and then click **Binary** or press the **F8** key).

We will now use the calculator to count in binary, starting at 1 and adding 1 each time. Click the "**1**" button and then the "+" button. Now click the "=" button to start counting. Just keep clicking it to count up by 1 each time – **figure 6.1**.

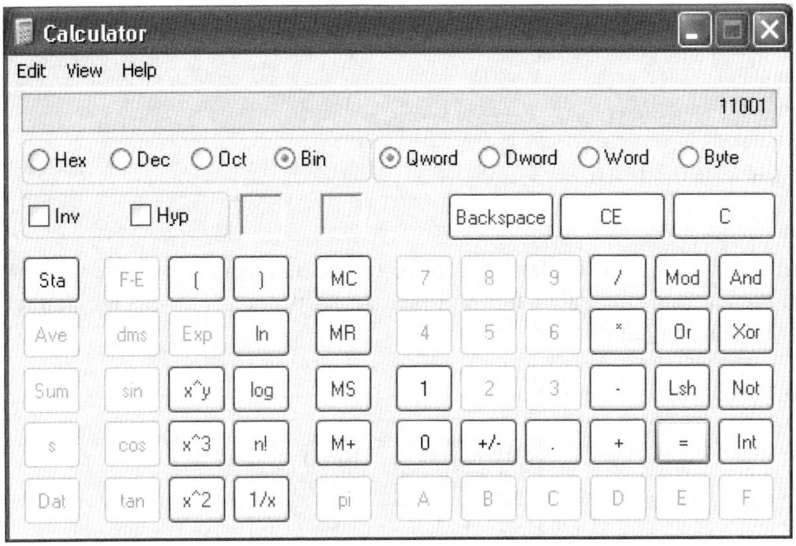

Figure 6.1: Windows Calculator Used for Counting in Binary

Any time that you would like to see what the equivalent decimal number is of your current binary number, just click the **Dec** radio button. Click the **Bin** radio button again to change back to binary, you can then continue to click the "=" button to carry on counting.

We have seen how decimal numbers are weighted. Binary numbers are weighted as follows:

$2^0 = 1$
$2^1 = 2$
$2^2 = 4\ (2 \times 2)$
$2^3 = 8\ (2 \times 2 \times 2)$
etc.

So the number 1110 when read from right to left consists of:
$0 \times 1 +$
$1 \times 2 +$
$1 \times 4 +$
1×8
$= 8 + 4 + 2 + 0$
$= 14$

Another example, the number 10100011
$1 \times 1 +$
$1 \times 2 +$
$0 \times 4 +$
$0 \times 8 +$
$0 \times 16 +$
$1 \times 32 +$
$0 \times 64 +$
1×128
$= 128 + 0 + 32 + 0 + 0 + 0 + 2 + 1$
$= 163$

You can use Windows Calculator to convert these numbers. First select binary mode, type the number into the calculator and then select decimal mode. The number will be converted when changing modes.

Table 6.1 shows a binary count of 0 to 15.

Decimal	Binary
0	0000
1	0001
2	0010
3	0011
4	0100
5	0101
6	0110
7	0111
8	1000
9	1001
10	1010
11	1011
12	1100
13	1101
14	1110
15	1111

Table 6.1: Decimal Numbers with Binary Equivalents

An easy way to learn how to write a sequence of binary numbers as seen in the table is to start with the right most bit (LSB) and write down a sequence of alternating numbers vertically starting from 0 i.e. 01010101... The next number to the left of this will consist of a sequence of numbers also starting from 0, but changing every two numbers i.e. 001100110011… To the left of this number, sequences of four numbers starting from 0 i.e. 0000111100001111… To the left of this number, sequences of eight numbers starting from 0 i.e. 0000000011111111… **Table 6.2** will make this clearer.

④ Start with 8 0s and change every 8 numbers

③ Start with 4 0s and change every 4 numbers

② Start with 2 0s and change every 2 numbers

① Start with 0 and change the number each time: 010101 etc.

MSB			LSB	Result
0	0	0	0	0000
0	0	0	1	0001
0	0	1	0	0010
0	0	1	1	0011
0	1	0	0	0100
0	1	0	1	0101
0	1	1	0	0110
0	1	1	1	0111
1	0	0	0	1000
1	0	0	1	1001
1	0	1	0	1010
1	0	1	1	1011
1	1	0	0	1100
1	1	0	1	1101
1	1	1	0	1110
1	1	1	1	1111

Table 6.2: An Easy Way to Count in Binary

Why we need to know about binary numbers

Embedded system programming consists of communicating with hardware devices attached to the microcontroller as well as hardware devices that are found on the microcontroller chip itself. These hardware devices are set up and operated by writing numbers to them and reading numbers back from them. Take the example of a microcontroller interfaced to four LEDs as in **figure 6.2**.

To switch on LEDs D1 and D4 we write the binary number 1001 to the LEDs. Because binary numbers are stored as voltage levels, we are actually connecting 5V to the anode of D1, 0V to the anode of D2, 0V to the anode of D3 and 5V to the anode of D4. When we write 1001 to the LEDs D1 and D4 will therefore switch on and LEDs D2 and D3 will switch off (or stay off).

We will see more examples of writing binary numbers to switch things on and off in the embedded programming chapters of this book. In C programs we actually use hexadecimal numbers instead of binary numbers, but hexadecimal numbers are a direct representation of binary numbers. You could also use decimal numbers in C as sending the number 9 to the LEDs will send the correct binary number of 1001, however it is easier to see what is being switched on or off when looking at the binary number. One can immediately see from the binary number that the first bit is switching something on, the middle two are switching something off and the last bit is switching something on.

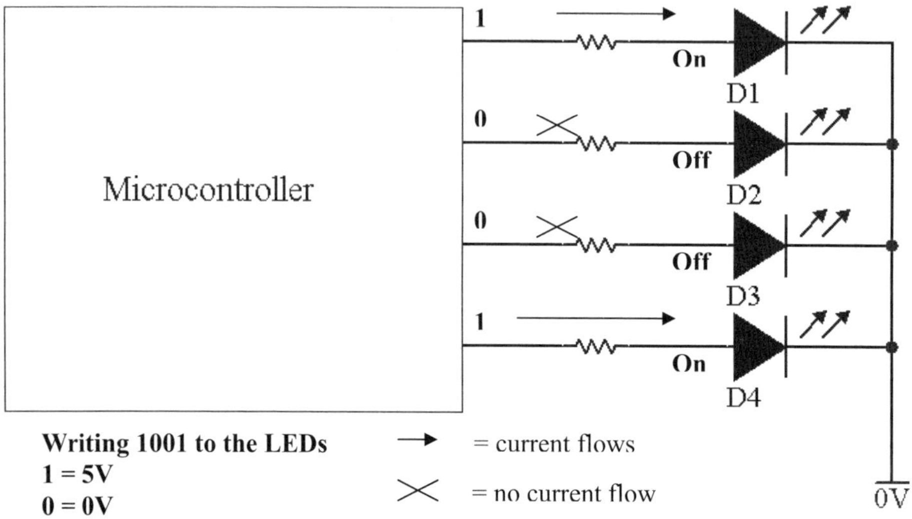

Figure 6.2: LEDs Interfaced to a Microcontroller

6.3.3 Hexadecimal Numbers

The hexadecimal number system is a shorter way to represent binary numbers. We can use hexadecimal numbers directly in our C programs. They make binary numbers easier to read.

We have seen the base 10 decimal system and the base 2 binary system. The hexadecimal system is base 16 (hex = 6, decimal = 10, hexadecimal = 6 + 10 = 16).

When counting in decimal, we use ten digits 0 to 9. In the hexadecimal system we have an extra 6 numbers to count to and use the letters A to F to represent them. So counting up in hexadecimal using a single digit will look like this:
0, 1, 2, 3, 4, 5, 6, 7, 8, 9, A, B, C, D, E, F

If we want to count higher than F, we are now out of digits and have to increment the place holder to the left of the single digit number, so the counting sequence continues as follows:
10, 11, 12, 13, 14, 15, 16, 17, 18, 19, 1A, 1B, 1C, 1D, 1E, 1F, 20, 21, 22, 23, 24, 25, 26, 27, 28, 29, 2A, 2B ...

Hexadecimal numbers are terminated with an H when written in text or with a 16 for example:
19A34C0H or $\mathbf{19A34C0}_{16}$
In C source code they are written starting with 0x (zero ex) as follows:
0x19A3C0

Each hexadecimal digit directly represents a four bit binary number. If you want to know how high you can count to using a binary number of x length, you use the formula:

max_count = 2^x - 1 (maximum count = two to the power of x minus one)

where **x** = number of bits in the binary number

If we want to know how high it is possible to count with just 2 bits, we calculate it as follows:

```
max_count = 2^x - 1
          = 2^2 - 1
          = 2 × 2 - 1
          = 4 - 1
          = 3
```

This means that it is possible to count to 3 using two binary digits. Let's check this:

00B = 0
01B = 1
10B = 2
11B = 3

With Windows Calculator in decimal mode, you can do this calculation more easily: press **2**, then press the **x^y** button, then press **2** and then =. You have now calculated **2^2**. Now subtract 1: press - and then **1** and finally =.

An 8 bit number can be used to count up to:

```
max_count = 2^x - 1
          = 2^8 - 1
          = 2 × 2 × 2 × 2 × 2 × 2 × 2 × 2 - 1
          = 256 - 1
          = 255
```

It is possible to count up to 255 with an 8 bit number.
Windows Calculator: Press **2**, **x^y**, **8**, -, **1**,=

Now let's look at a 4 bit number:

```
max_count = 2^x - 1
          = 2^4 - 1
          = 2 × 2 × 2 × 2 - 1
          = 16 - 1
          = 15
```

Windows Calculator: Press **2**, **x^y**, **4**, -, **1**, =
15 is the same number that a single hexadecimal number can count up to (F hex = 15 dec). Note that there are 16 hexadecimal numbers, but the first number is zero. Counting from 1 to F in hexadecimal is the same as counting from 1 to 15 in decimal or 1 to 1111 in binary (**table 6.3**).

You can use Windows Calculator again to convert between numbers from different number systems. Switch the calculator into whichever number system you want to convert from by using one of the radio buttons Hex for hexadecimal, Dec for decimal or Bin for binary. Type the number in that you want to convert and then using the radio buttons again, select the number system that you would like to convert to.

To count up in hexadecimal, put the calculator into hexadecimal mode. Type "**1**" and then "+" and then continually press the "=" key to count up.

Decimal	Hex	Binary
0	0	0000
1	1	0001
2	2	0010
3	3	0011
4	4	0100
5	5	0101
6	6	0110
7	7	0111
8	8	1000
9	9	1001
10	A	1010
11	B	1011
12	C	1100
13	D	1101
14	E	1110
15	F	1111

Table 6.3: Counting in Decimal, Hexadecimal and Binary

Why we need the hexadecimal number system in C programming

So that we can switch external hardware devices on and off just like with the binary example. You can not type a binary number directly into your C source code to use it in your program. You could write a function to do this, but an easier way is to use a hexadecimal number. Hexadecimal numbers are also shorter than binary when typed. For example the 32 bit binary number:

10000101111100001111000010101100

is written in hexadecimal as:

0x85F0F0AC

How to use hexadecimal numbers in C

If you had a microcontroller with a 32 bit port connected to 32 LEDs, you could write the above number to the port to switch the LEDs on and off. Let's take a simpler example of an 8 bit port with 8 LEDs interfaced to it.

Let's switch the first two LEDs on, the next two off, the next 3 off and the last one on (2 on, 2 off, 3 off 1 on). Notice firstly that I am visualising the LEDs in groups of four (groups of 1 hexadecimal digit each) that is why I have not written 2 on, 5 off, 1 on). The binary number to switch this sequence will be:

1100 0001

Now convert it to hexadecimal to use it in a C program. You can just look up the numbers in **table 6.3** or type it into Windows Calculator and convert it. When you have worked with binary and hexadecimal numbers for a long time, you will be able to convert it in your head. When using the table to look up the hexadecimal number, group the binary number into four bits each starting from the right of the number and look up each four bit number in the table.

0001B is **1H**

1100B is **CH**

The number will therefore be:

C1H

or

0xC1

when written in your C program.

How to convert from binary to hexadecimal in your head

You might want to come back to this later after you have worked with bin and hex numbers for a while, but here it is anyway.

You always work in groups of four digits, starting from the right of the binary number. Each bit is weighted as follows:

Weight	8	4	2	1
Bit number	3	2	1	0

Note that when numbering bits, they are numbered from right to left starting at 0.

Now apply the weighting to each group of four bits by multiplying the weight by the binary digit under it. Add the results of this operation together, and then count in hex to convert. A few examples will clarify.

To convert the previously used 8 bit number **1100 0001** to hexadecimal:

Taking the lower nibble (first four bits from the right: 0001) line it up under the weight values.

Weight	8	4	2	1
Number	0	0	0	1

Now multiply each bit by its weight and then add the results:

```
8 × 0  +  4 × 0  +  2 × 0  +  1 × 1  =  1
```

Now the upper nibble (left four bits of the byte):

Weight	8	4	2	1
Number	1	1	0	0

```
8 × 1  +  4 × 1  +  2 × 0  +  1 × 0 = 12
```

The number 12 is in decimal, so we need to count through this number to convert it to hexadecimal. Start counting 1, 2, 3, 4, 5, 6, 7, 8, 9, we can't count ten we must count A, then B, C that's 12 counts so 12 dec = C hex. Now put the two numbers together in the correct order and you get: 0xC1

Another example using the 32 bit number shown earlier:
`10000101111100001111000010101100`

First break it up into groups of 4:
`1000 0101 1111 0000 1111 0000 1010 1100`

Now convert each group of four. You may convert the groups from left to right, so long as you have broken them up from right to left. This is because if you have say a 7 bit number like `1010111`, it is broken up as `101 0111` and not `1010 111`.

I will convert each group from left to right for this example.
`1000 = 1 × 8 + 0 × 4 + 0 × 2 + 0 × 1 = 8`
No need to count because 8 dec = 8 hex

`0101 = 0 × 8 + 1 × 4 + 0 × 2 + 1 × 1 = 5`
5 dec = 5 hex
Start assembling the number: 0x85

`1111 = 1 × 8 + 1 × 4 + 1 × 2 + 1 × 1 = 15`
Start counting to 15 by counting through the hex number sequence. After 9, count A, B, C, D, E, **F**. You will quickly learn to recognise that when all 4 bits of the number are 1s, the biggest hex digit is used i.e. F.
The number is now 0x85F

`0000 = 0 × 8 + 0 × 4 + 0 × 2 + 0 × 1 = 0`
0 again
The number is now 0x85F0, we are half way there…

`1111 = 1 × 8 + 1 × 4 + 1 × 2 + 1 × 1 = 15`
F again.
The number is now 0x85F0F

`0000 = 0 × 8 + 0 × 4 + 0 × 2 + 0 × 1 = 0`
0 again.
The number is now 0x85F0F0

`1010 = 1 × 8 + 0 × 4 + 1 × 2 + 0 × 1 = 10`
Start counting to ten in hex. After 9 comes ten, but in hex it is **A**.
The number is now 0x85F0F0A. Last number coming up…

`1100 = 1 × 8 + 1 × 4 + 0 × 2 + 0 × 1 = 12`
Count to 12 in hex 1…9, A, B, **C**.

The number is **0x85F0F0AC**.

Let's look at one final example, but this time simplifying things a bit. We will line up our 8421 weighting above each 4 bit section of the number in our minds and whenever there is a 1 in the number, we will write the weight down. Whenever there is a 0 in the number, we will write 0 down. Our example number is:
11011110

Split it up into fours:
1101 1110

Now do the conversion:
1101 = 8 + 4 + 0 + 1 = 13
Count up to 13 in hex: 1…9, A, B, C, **D**.

1110 = 8 + 4 + 2 + 0 = 14
Count up to 14 in hex: 1…9, A, B, C, D, **E**.

The number is therefore: **0xDE**

6.4 Working with Hexadecimal Numbers in C

Now that we have seen how hexadecimal and binary numbers work, let's see how to use them in some C programs. You won't be able to use a binary number directly in your C program, but will use the hex equivalent instead.

The **printhex.c** program will demonstrate how to print a hexadecimal number:

```
#include <stdio.h>
#include <conio.h>

int main(void)
{
    printf("%x", 0xA914);         // print a hex number
    printf("\n0x%x", 0xA914);     // print same number with 0x
    printf("\n0x%X", 0xA914);     // hex number in capitals

    getch();
    return 0;
}
```

The example program shows that to write a hexadecimal number in C, it is always started with 0x. When printing a number in hexadecimal using **printf()**, the **%x** format specifier is used. The second time that the number is printed out, 0x is included in the text string so that the hexadecimal number is displayed as 0xa914 rather than a914. The third time that the hexadecimal number is printed out, the format specifier uses a capital x (**%X**).

This prints the hexadecimal number using capital letters for the numbers A to F: 0xA914 instead of 0xa914.

Numbers in a computer are always stored as binary numbers. When we print out a number in decimal or use a decimal number in our program, we are only displaying the number on the screen as a decimal number. We can view the same number as a hexadecimal number. When typing a decimal or hexadecimal number into a source code file, we are using a number that is easy for people to read. After compiling the source code and then loading the program to memory to run it, all of the numbers from the source code are converted to and stored as binary numbers. The next program **printnum.c** will demonstrate that we can view the same number stored in a variable as either decimal or hexadecimal.

```
#include <stdio.h>
#include <conio.h>

int main(void)
{
    int the_num;

    the_num = 498;
    printf("The number is %d in dec\n", the_num);
    printf("The number is %X in hex\n", the_num);
    // change the number
    the_num = 0xFA39C;
    printf("\nThe number is %d in dec\n", the_num);
    printf("The number is %X in hex\n", the_num);

    getch();
    return 0;
}
```

When the variable **the_num** in this program is assigned the value of 498, it is actually being assigned the binary equivalent of the decimal number 498 which is 111110010B (check this on your Windows Calculator). In reality, a location in memory is set aside for this variable and after assigning it this number, it contains the voltage levels:

5V, 5V, 5V, 5V, 5V, 0V, 0V, 5V, 0V

Where 5V represents whatever voltage level is used on the system for the binary value 1. Whether we choose to display the number as hexadecimal or decimal in our source code or screen makes no difference, the number is always stored as a binary number on the computer, consisting of voltage levels.

And now for a program that converts hexadecimal to decimal and vice versa, **convnum.c**.

```
#include <stdio.h>
#include <conio.h>

void menu(void);

int main(void)
```

```
{
    char menu_item = 0;
    int num;

    while (menu_item != 'x') {
        menu();                         // display the menu
        menu_item = getch();            // get key press
        if (menu_item == 'h') {
            printf("\nEnter hex num: ");
            scanf("%x", &num);
            printf("\nNumber in dec is %d\n", num);
        }
        else if (menu_item == 'd') {
            printf("\nEnter dec num: ");
            scanf("%d", &num);
            printf("\nNumber in hex is %X\n", num);
        }
        else if (menu_item != 'x') {
            printf("\nInvalid menu selection...\n");
        }
    }
    return 0;
}

void menu(void)
{
    printf("\nPress h to convert hex to dec.");
    printf("\nPress d to convert dec to hex.");
    printf("\nPress x to exit.\n");
}
```

A function is used for displaying a menu to the user in this program. The program waits for the user to select one of the menu items and then, depending on the user's selection, will either prompt the user to convert a number from hexadecimal to decimal or decimal to hexadecimal or exit the program. When converting the hexadecimal number to decimal, the number can be entered in upper or lowercase, with or without a preceding 0x e.g. f1ca, F1CA, 0xF1CA. As in the previous program, the number stored in a variable is printed in either hexadecimal or decimal to the screen.

Notice the last **else if** statement in the program. A check is done to see that the user has not selected 'x'. This statement will evaluate to true when the user selects anything except 'h', 'd' and 'x'. 'h' and 'd' because if they are selected will be taken care of by the previous two decisions and 'x' because we are now selecting any remaining possible characters besides 'h', 'd' and 'x'. If 'x' is selected, the program will do nothing until the loop expression is evaluated again and then skip over the loop and exit the program.

6.4.1 Field Width Specifiers Revisited

We have already seen field width specifiers used with floating point numbers to set the number of digits to print after the decimal point. It is also possible to set the number of digits before the decimal point, or in the case of decimal and hexadecimal numbers, set the number of place holders to the left of the number. As an embedded programmer, you may wish to use this to show the width of a number when using hexadecimal numbers, or to align numbers. For example, the number 0x5 is a 3 bit number (101_2), but in a computer, memory is always displayed in multiples of bytes, so we may wish to display this number with padding zeros to represent all 8 bits – 0x05 ($0000\ 0101_2$). The next program, **padding.c**, shows how to pad different numbers with zeros or spaces.

```
#include <stdio.h>
#include <conio.h>

int main(void)
{
    // a 3 bit hex number padded to 8 bits
    printf("0x%X\n", 0x5);          // unpadded
    printf("0x%2X\n", 0x5);         // pad with spaces
    printf("0x%02X\n\n", 0x5);      // pad with 0

    // a 3 bit hex number padded to 16 bits
    printf("0x%4X\n", 0x5);         // pad with spaces
    printf("0x%04X\n\n", 0x5);      // pad with 0

    // decimal number padded
    printf("%d\n", 123);            // unpadded
    printf("%5d\n", 123);           // padded with spaces
    printf("%05d\n\n", 123);        // padded with 0

    // float padded
    printf("%f\n", 34.218);         // unpadded
    printf("%10f\n", 34.218);       // padded with spaces
    printf("%10.4f\n", 34.218);     // pad = space, 4 nums after .
    printf("%010.4f\n\n", 34.218);  // pad = 0, 4 nums after .

    getch();
    return 0;
}
```

Run the example program and note the following:

1. A number placed in the middle of the format specifier (directly after the %) is the field width specifier (which is now part of the format specifier).

2. A field width specifier will pad a number with spaces unless it is preceded with 0, in which case it will pad the number with 0s.

3. The value of the field width specifier is the total width of the number printed. So if the number is 23 and the field width specifier is 1 or 2, you will not see any padding. As soon as a width of 3 or more is used, the padding will be seen as the number is forced to a length of 3 digits instead of 2 in this example.

4. When using field width specifiers with floating point numbers, the number in front of the dot of the field width specifier specifies the entire length of the number printed including the decimal point and all numbers in front of and behind it. The number

after the dot in the field width specifier specifies the number of digits after the decimal point in the printed number.

Figure 6.3 shows the above points diagrammatically.

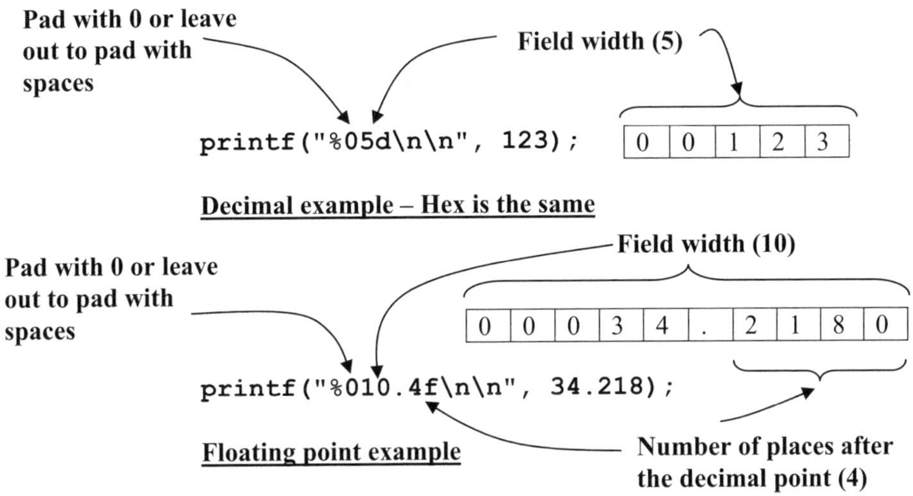

Figure 6.3: Field Width Specifiers in Decimal and Floating Point Numbers

6.5 The ASCII Alphanumeric Code

You have been using Programmer's Notepad or some other ASCII text editor to write your C programs. When working with character data types (char) in C programs, you are working with ASCII characters. What does this mean?

As already mentioned, computers only work with numbers. If I want to send text from one computer to another or write and save text to a computer, I can only use numbers to do this. When a text file is opened, the computer knows that the file is a text file as it has a **`.txt`** file extension or other file extension that lets the computer know that the file contains text. When the computer reads the contents of the text file and displays it on the screen, it will read a byte at a time and will interpret each byte. If it reads, for example, the number 0x41 (01000001B or 65), it will interpret this and display the letter 'A' on the screen. The computer uses the ASCII code to look up this number and display the correct letter or symbol.

ASCII stands for "American Standard Code for Information Interchange". The ASCII code consists of letters from the alphabet, numbers 0 to 9 and other characters like brackets and punctuation marks. When the computer reads ASCII text, it will look up the numbers that the text represents in the ASCII code and display the corresponding letter or symbol. The original ASCII table consisted of 7 bit numbers (0 to 127), but it has been extended to 8 bit numbers (0 to 255) to support languages other than English as well as add some graphical symbols. See **appendix A** for the extended ASCII table.

The next program **asciiread.c** will use functions from the C standard library to open a text file, read the file and display the contents as characters, hexadecimal numbers and decimal numbers. Before running this program, you need a text file called **textfile.txt** to be present in the directory in which your compiled program exists (the program will read this file). You can use Programmer's Notepad to create this file. Save the following line of text to it:
A text string 12345.

```
// asciiread.c
#include <stdio.h>
#include <conio.h>

int main(void)
{
    FILE *fptr;       // pointer to file
    char chr;         // stores character read from file

    fptr = fopen("textfile.txt", "r");  // open file
    while ((chr = getc(fptr)) != EOF) {
        printf("%c\t0x%02x\t%d\n", chr, chr, chr);
    }
    fclose(fptr);     // close file
    getch();
    return 0;
}
```

The program opens the **textfile.txt** file and then displays each character in the file as a character, as a hexadecimal number and as a decimal number as follows:

```
A       0x41    65
        0x20    32
t       0x74    116
e       0x65    101
x       0x78    120
t       0x74    116
        0x20    32
s       0x73    115
t       0x74    116
r       0x72    114
i       0x69    105
n       0x6e    110
g       0x67    103
        0x20    32
1       0x31    49
2       0x32    50
3       0x33    51
4       0x34    52
5       0x35    53
```

You can look up each of the alphanumeric values in the left column in the ASCII table in **appendix A** and see that the numbers to the right correspond correctly.

The functions **`fopen()`** and **`fclose()`** are C standard library functions for opening and closing files. We need a file pointer to be able to use these functions. The line:
`FILE *fptr;`
creates a pointer to a file structure. We will cover pointers in the next chapter and structures in a later chapter, so just accept that we need to do this for now. The structure **`FILE`** is defined in the **`stdio.h`** header file. When opening the text file, we pass **`"r"`** to the **`fopen()`** function. This opens the file for reading or "read only", not for writing to.

In the **`while`** loop's loop expression, we are calling the **`getc()`** function, another C standard library function, to get a character from the file that has been opened. The character is assigned to the character variable **`chr`**. A check is also done in the same line to see if this character is not equal to **`EOF`**. **`EOF`** is defined in the **`stdio.h`** file using **`#define`** and stands for "End of file". By checking for this value, we will know if we have reached the end of the file and can exit the loop. This value is not a character, it is a value that the operating system sends to the **`getc()`** function when the end of the file has been reached. When **`getc()`** returns this value, the while loop will be exited. After each pass through the loop, the **`getc()`** function is run and the check for **`EOF`** is done. Each time that **`getc()`** is run, it gets and returns the next character from the file. The first time that it is called, it will get the first character from the file, thereafter, it will get each successive character from the file.

Inside the **`while`** loop, the **`printf()`** function prints out the character from the file in three different ways – as a character, as a hexadecimal number and as a decimal number. There is one new thing to you here, the **`\t`** escape sequence. This escape sequence inserts the **`Tab`** character into the text string to be printed. This results in the printed values each having their own column. Let's examine the printed string:

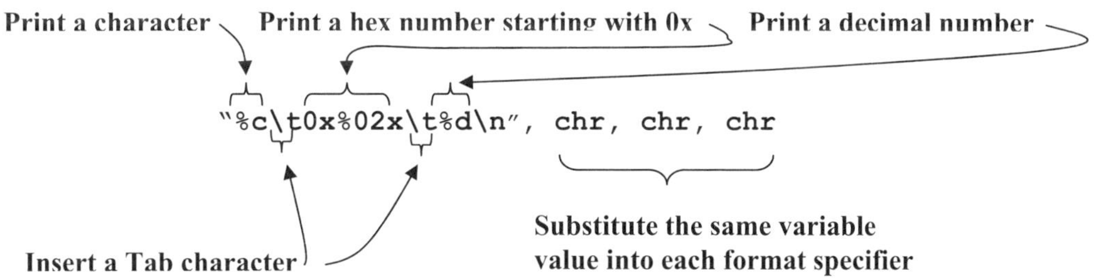

From this chapter I hope that you will understand that:
Numbers are stored as voltage levels on a computer:

```
0V, 5V, 0V, 0V, 0V, 0V, 5V, 5V
```

Note that when copying data from magnetic or optical media to memory, data is changed from being stored as magnetic fields or pits and lands (CDs) to voltage levels. The computer can only work with voltage levels, so all data that must be processed, no matter where it is stored, is first changed to voltage levels and later if saved to other media, changed back to magnetic fields or however it is stored there.

These voltage levels are known as "logic levels" and represented as binary numbers:

```
0100 0011
```

These same logic levels or binary numbers can be viewed by people or written in C programs as decimal numbers:

```
67
```

Or they can be viewed as hexadecimal numbers:

```
0x43
```

Or they can be viewed as characters:

```
C
```

So whether you type 'C', 67 or 0x43 into your C source file, you are essentially typing the same binary number of 01000011 and can choose in which number system to display it on the screen when your program prints it out. Note that this is not specific to the C language, but to computers and digital systems in general.

One last example, **asciiwrite.c** will show that creating a file in a C program and writing numbers to the file from the ASCII table will result in the file containing text.

```
// asciiwrite.c
#include <stdio.h>
#include <conio.h>

int main(void)
{
    FILE *fptr;

    // create a text file to write to
    fptr = fopen("asciiout.txt", "w");
    putc(0x48, fptr);    // write bytes to the text file
    putc(0x45, fptr);
    putc(0x4C, fptr);
    putc(0x4C, fptr);
    putc(0x4F, fptr);
    fclose(fptr);        // close the file

    printf("Finished writing file: asciiout.txt");
    getch();
```

```
    return 0;
}
```

This program creates a new file by passing **"w"** (write) to the **fopen()** function. A file **asciiout.txt** is created in the directory in which this program is run and is opened for writing to. Some hexadecimal numbers are then written to the file. You can look up these numbers in the ASCII table in Appendix A and see that they spell out the word "HELLO". When this program is run, the **asciiout.txt** file is created, and when you open this file you will see that the file contains the text "HELLO". The program that you have used to view the file is interpreting the numbers using the ASCII table and this is why you see this text. 0x48 = H, 0x45 = E, etc.

6.6 Exercises

1. A microcontroller is interfaced to an 8 bit port that is driving 8 LEDs. What hexadecimal number would you write to this port to switch the first two LEDs on (left most LEDs), the second two LEDs off, the next 3 LEDs on and the last (right most) LED off?

2. Write a program that uses a field width specifier to print the following decimal number in hexadecimal format with a bit width of 16 bits in a C program: 2779. Fill the unused place holders with zeros.

3. Write a program to print the hexadecimal number 0x5f as a character, as a decimal number and as a hexadecimal number.

6.6.1 Solutions

1. The LEDs will be On, On, Off, Off, On, On, On, Off. This equals:
   1100 1110B
   The hexadecimal equivalent will be:
   8 + 4 for the first hex digit on the left (left four bits)
   = a count of 12
   = count from 10 to 12 in letters starting from A
   = C
   8 + 4 + 2 for the hex digit on the right (right four bits)
   = a count of 14
   = count from 10 to 14 in letters starting from A
   = E
   Put the two together and the answer is:
   0xCE

2. 16 bits is equal to 4 hexadecimal digits (4 bits each)
   printf("%04X", 2779);

3. **printf("%c\t%d\t%X", 0x5F, 0x5F, 0x5F);**

6.7 Summary

- Numbers are stored as voltage levels in a computer's memory.
- These voltage levels are represented in binary as zeros (0s) and ones (1s).
- Writing a number to an external port on a computer can switch devices that are attached to the port on or off.
- Reading a number from an external port will get the state of external devices that are attached to the port. The state of these devices will be reflected in the number read from the port.
- Binary numbers can be represented conveniently and in a more compact format by hexadecimal numbers.
- In the C programming language, hexadecimal numbers can be used directly to represent binary numbers and are written staring with 0x.
- The ASCII alphanumeric code is used to interpret numbers as characters.
- Any number in a computer or microcontroller's memory can be represented as a binary number, a hexadecimal number and a decimal number. If the number is read as an 8 bit number it can be represented as an ASCII character.

7 Memory and Microcontrollers

What you will do in this chapter:

- Find out how memory works
- Learn how a microprocessor accesses memory and peripherals
- See how to access memory by using the address of the memory
- Learn about other data types in C
- Choose your microcontroller and embedded system

7.1 Memory Basics

Two main types of memory are present on microcontrollers namely ROM and RAM.

ROM (Read Only Memory): This type of memory is used to store the program that we write and compile. It is *non-volatile* memory meaning that it will retain its contents even if the power is switched off. ROM on a modern microcontroller is usually present as Flash memory. Flash memory can be programmed while it is still in place on the circuit board. The contents of the Flash memory are normally not altered when a user application program is running, although some microcontroller's Flash memory is in-application programmable. The program stored in ROM memory will be the program that is run when power is supplied to the microcontroller.

RAM (Random Access Memory): RAM is used to store variables and data from our program that can be changed while the program is running. This type of memory is *volatile* meaning the contents of the chip will be lost when power is removed. Microcontrollers usually contain some SRAM (Static RAM).

Data is stored in memory as an array of bytes. Each byte has a unique address and can be individually addressed. Data is stored to memory by a write operation and retrieved from memory by a read operation. Each bit in a byte of the memory chip is called a cell and can hold either a 1 or a 0 logic level as shown in **figure 7.1**. The capacity of the memory chip is determined by how many addressable bytes it contains.

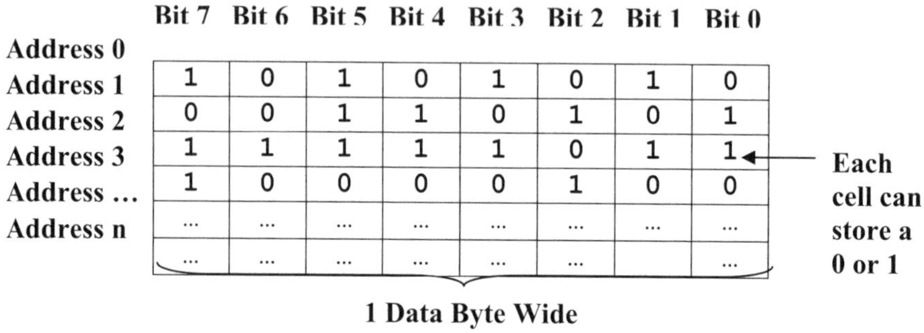

Figure 7.1: Memory Showing How Data is Stored

7.2 A Look at a Memory Chip

Let's take a look at how a SRAM memory chip works in order to better understand memory. **Figure 7.2** shows the top view of a 2kB (two kilo byte) memory chip. **Figure 7.3** shows a block diagram of the same chip. The size of the memory is determined by how many bytes it can address. This is determined by the number of address lines on the chip. The chip has 11 address lines and so can address 2^{11} bytes = 2048 bytes (or 2 kilo bytes because 1 kilo byte = 1024 bytes (2^{10})).

A bus is just a group of wires on a computer system. The address bus is the group of wires that are used for addressing memory and peripheral devices. It is 11 bits wide on the memory chip shown. The data bus is the group of wires used for sending and receiving data to and from memory and peripheral devices. The data bus on the memory chip in the figure is 8 bits wide. The control bus is the group of wires used to control a memory chip or peripheral device. There are 3 control lines on the memory chip. A peripheral device is any piece of hardware interfaced to the microcontroller, whether it is an internal device or external to the microcontroller.

The procedure for writing data to or reading data from the memory chip is described below. If you plugged this memory chip into an electronic breadboard and used wires to connect the chip as described, you could write data to the chip and read data from the chip by plugging the correct sequences of wires to the correct power supply terminals and reading the data by measuring voltage levels with a voltmeter.

In order to operate the memory chip of **figure 7.2**, power must first be supplied to the chip on the VDD and GND pins – the positive voltage specified in the chips datasheet is supplied to VDD from a power supply and the 0V of the power supply is connected to the GND pin. The three control lines of the control bus, CS, OE and WE must be "pulled high" i.e. they must have a logic 1 applied to them – connected to the positive of the power supply. This is because they are "active low" meaning that they are active when logic 0 is applied to them. We know this because of the line above these names as shown in the figure. If there was no line, they would be "active high" – activated by a logic 1 level. The chip will now be in an inactive state and the voltage levels measured on the address and data bus will be "high impedance" meaning that they are neither at logic 0 nor at logic 1, but are offering a high resistance – more like an insulator than a conductor.

From this inactive state, we can write something to the chip. First we decide at what address we would like to save the data byte that we are going to write. If we choose the first address in the chip, address 0, then we must supply the address 000 0000 0000B to the address bus. In other words we must connect the 11 address lines of the address bus (A0 to A10) to 0V of the chips power supply. Remember that we are counting from 0 so A0 to A10 is 11 lines, not 10.

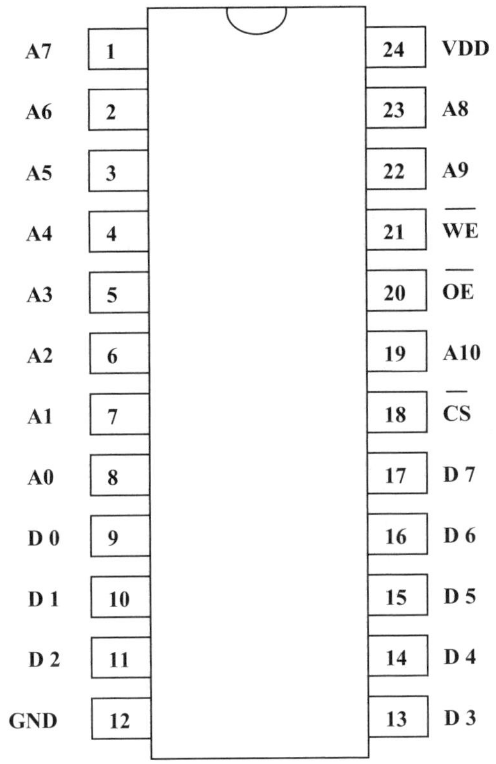

VDD and GND: **Power supply to chip**
A0 to A10: **Address bus – data byte address placed here**
D0 to D7: **Data bus – 8 bit data byte read and written here**
$\overline{CS}$: **Chip select**
$\overline{OE}$: **Output enable** } **Control bus**
WE: **Write enable**

Figure 7.2: An 8 Bit SRAM Memory Chip

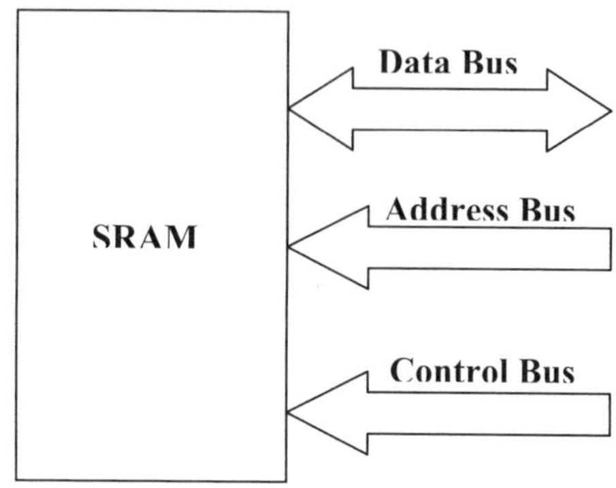

Figure 7.3: Block Diagram of an SRAM Memory Chip

Next we must decide what data byte to write. Let's say 120 decimal. This is the equivalent of 78 in hexadecimal which is 0111 1000 in binary. We must therefore connect D7 of the data bus to 0V of the power supply, D6 to D3 to the positive of the power supply and D2 to D0 to the 0V of the power supply.

The address is now addressing the first data byte in the memory chip, the data bus has the data on it that we would like to save to the chip, we must now operate the control signals to move the data into the chip. The chip must first be selected by putting logic 0 on the CS pin i.e. connect it to the 0V of the power supply. Next we put logic 0 on the WE pin and this moves our data byte into the chip and saves it at address 0. The CS and WE pins can now be moved back to logic 1 to disable the chip again. The chip will now keep our data byte stored at address 0 until we change it by writing a byte to the same address or the power is switched off. This whole operation is known as a write operation.

To read our data byte back from the chip we must first apply the address to the chips address bus again. No voltage levels must be applied to the data bus or they may be shorted out when the read is performed. Next the chip is enabled with the CS pin pulled to logic 0. The output enable (OE) pin is now pulled to logic 0 and the byte that we saved at address 0 will now appear as binary voltage levels on the data bus. You would be able to measure them with a voltmeter.

This same process can be applied to any data byte in the chip by changing the address on the address bus. Want to read or write byte number 1045? Apply the address 1045 to the address bus – 0x415 or 100 0001 0101B and then follow the read or write procedure.

Figure 7.3 shows the write and **figure 7.4** shows the read of the memory chip graphically.

Address Bus												Data Bus									Control Bus		
A10	A9	A8	A7	A6	A5	A4	A3	A2	A1	A0		A7	A6	A5	A4	A3	A2	A1	A0		CS	WE	OE
Z	Z	Z	Z	Z	Z	Z	Z	Z	Z	Z		Z	Z	Z	Z	Z	Z	Z	Z		1	1	1
0	0	0	0	0	0	0	0	0	0	0		Z	Z	Z	Z	Z	Z	Z	Z		1	1	1
0	0	0	0	0	0	0	0	0	0	0		0	1	1	1	1	0	0	0		1	1	1
0	0	0	0	0	0	0	0	0	0	0		0	1	1	1	1	0	0	0		0	1	1
0	0	0	0	0	0	0	0	0	0	0		0	1	1	1	1	0	0	0		0	0	1
Z	Z	Z	Z	Z	Z	Z	Z	Z	Z	Z		Z	Z	Z	Z	Z	Z	Z	Z		1	1	1

Rows from top to bottom:

Step 1:Chip is inactive (Z = high impedance or disconnected)

Step 2:Address is applied to address bus

Step 3:Data to be written is applied to the data bus

Step 4:Chip select CS is enabled

Step 5:Activating the WE control line moves the data into the chip at the selected address

Step 6:Chip is put back into the inactive state

Figure 7.3: Writing 0x78 to a Memory Chip at Address 0

Address Bus												Data Bus									Control Bus		
A10	A9	A8	A7	A6	A5	A4	A3	A2	A1	A0		A7	A6	A5	A4	A3	A2	A1	A0		CS	WE	OE
Z	Z	Z	Z	Z	Z	Z	Z	Z	Z	Z		Z	Z	Z	Z	Z	Z	Z	Z		1	1	1
0	0	0	0	0	0	0	0	0	0	0		Z	Z	Z	Z	Z	Z	Z	Z		1	1	1
0	0	0	0	0	0	0	0	0	0	0		Z	Z	Z	Z	Z	Z	Z	Z		0	1	1
0	0	0	0	0	0	0	0	0	0	0		Z	Z	Z	Z	Z	Z	Z	Z		0	1	0
0	0	0	0	0	0	0	0	0	0	0		0	1	1	1	1	0	0	0		0	1	0
Z	Z	Z	Z	Z	Z	Z	Z	Z	Z	Z		Z	Z	Z	Z	Z	Z	Z	Z		1	1	1

Rows from top to bottom:

Step 1:Chip is inactive (Z = high impedance or disconnected)

Step 2:Address is applied to address bus

Step 3:Chip is selected by activating CS

Step 4: Output enable OE control line is activated

Step 5: Data being read now appears on the data bus

Step 6:Chip is put back into the inactive state

Figure 7.4: Reading Data from a Memory Chip at Address 0

7.3 How Microprocessors Access Memory and Peripherals

A microprocessor has an address bus, a data bus and a control bus. The memory chip is interfaced to the microprocessor by connecting its busses to the microprocessors busses. On a microcontroller with on-chip memory, this is already done inside the chip.

The microprocessor contains registers – these are similar to memory locations in the microprocessor as they store a number. One of these registers is called the program counter (PC) and the value in this register is placed on the microprocessors address bus. When power is applied to the microprocessor, the program counter is reset to the value of 0. So at reset, an address of 0 appears on the microprocessors address bus (all address lines are at logic 0). When a non-volatile memory chip is wired to the microprocessor at this address and it contains a program, the microprocessor will automatically fetch the first instruction of the program from the memory chip at address 0 and execute it (an instruction is just a binary number). The program counter will then count up by one and fetch the instruction from the memory at address 1 and execute it. The program counter will keep counting up and fetching each successive byte of the program from the memory chip and execute it. If a jump instruction occurs, the address in the program to jump to is loaded into the program counter and execution of instructions continues from the new address with the program counter incrementing the address after each instruction is fetched.

This is how a microprocessor or microcontroller will "boot up" – by starting to run a program at a known address. The start up address does not have to be 0, but it usually is. Some microprocessor manufacturers will design their chips to start running at a different address. The non-volatile chip containing the program to be run must then be wired to this address by the hardware engineer who builds the system or by the microcontroller manufacturer when the microcontroller is designed.

A peripheral device, like a bank of 8 LEDs, can be interfaced to the microprocessor by using a buffer chip that is like a single memory location. The chip will hold the voltage levels written to it and will be able to deliver enough current to drive the LEDs (switch them on). The voltage levels that the chip is holding are available externally on the pins of the chip and this is where the LEDs are connected. So writing to a single memory location in a memory chip and writing to a bank of LEDs is the same operation in a microprocessor. Writing to the memory location just saves a number, writing to the LEDs displays the number on the LEDs in binary where an LED that is on represents binary 1 and an LED that is off represents a binary 0. **Figure 7.5** shows a block diagram of a microprocessor interfaced to Flash memory, SRAM memory and peripheral devices.

In C, we can write to or read from a peripheral that is at a specific address. This will cause the microprocessor to load this address on the address bus and fetch data from or send data to the device on the data bus. Addresses in C can be accessed by pointers. We will look at these next.

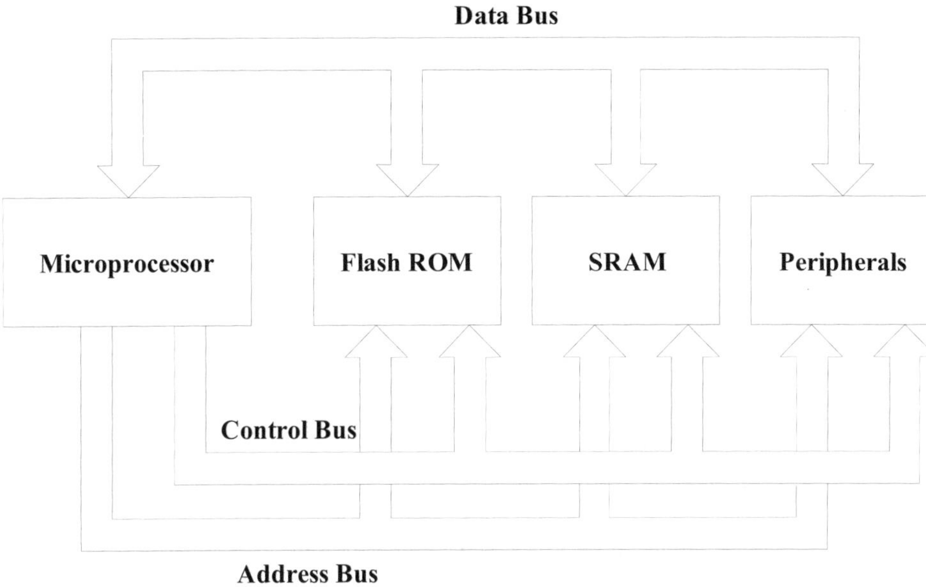

Figure 7.5: Microprocessor Interfaced to Memory and Peripherals

7.4 Pointers

Pointers in C point to a location in memory or to control registers of a peripheral device. They contain the address of the device or memory location. Before we start to use pointers, let's write a program that displays the address of a variable in memory. We know that a variable is a piece of memory set aside by the compiler for use in a program. Because the variable exists in memory, it must have an address. The **address.c** program displays the address of an integer variable.

```
/* address.c */
#include <stdio.h>
#include <conio.h>

int main(void)
{
    int var = 100;

    printf("The variable var contains the value %d\n", var);
    printf("The address of var is %d", &var);

    getch();
    return 0;
}
```

In this program, the value that the variable **var** is holding is printed first. The address of this variable is then printed out. To print the address of the variable, we use the ampersand character (**&**). This should be familiar to you from using the **scanf()** function where we passed the address of the variable in which we wanted to save data entered by the user to **scanf()**. When using the name of a variable in C we are referring to the value that the variable (or memory location) is storing. When preceding the variable name with an ampersand, we are referring to the address of the variable in memory. See **figure 7.6**.

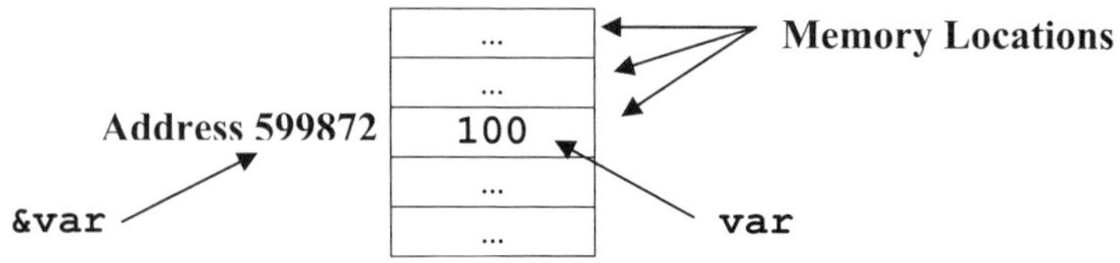

Figure 7.6: Variable Address and Variable Data in C

For our next example, we create a pointer and make it point to a variable. If the variable was a port with LEDs attached to it, we would then be able to control the LEDs by writing to them.

```
/* led_port.c */
#include <stdio.h>
#include <conio.h>

int main(void)
{
    int LEDs = 0x88;        /* a variable representing LEDs */
    int *ptr;               /* a pointer */

    ptr = &LEDs;            /* get the address of the LEDs */
    printf("LEDS are set to %x\n", *ptr);
    *ptr = 0x44;
    printf("LEDS are set to %x", *ptr);

    getch();
    return 0;
}
```

In an embedded system the LED port will be at a specified address. The hardware engineer who designed the system would have allocated it to the address, and he would have to specify that the port is at, for example, address 0xFFFF 0500. On the PC, when a variable is defined, we don't have any control of the exact location in memory that the compiler will put it, so in the above program, the address of the variable is obtained when the program is running. Let's see how it works.

A pointer variable is defined near the top of the body of **main()**:
int *ptr;
This tells the compiler that we want to store the address of an integer in this pointer variable. The next line assigns the address of the variable **LEDs** to the pointer variable.

We can now use the pointer to access the contents of the variable **LEDs**. Putting an asterisk (*) in front of the pointer variables name dereferences the pointer, in other words it accesses the data that the pointer is pointing to.

This example may not look very impressive as you could just as easily use the variable **LEDs** to access the contents of this variable, but in an embedded system, the first few lines of the program:

```
int LEDs = 0x88;     /* a variable representing LEDs */
int *ptr;            /* a pointer */
ptr = &LEDs;         /* get the address of the LEDs */
```

would be replaced by assigning a specific address to the pointer e.g. 0xFFFFF430. You would now use ***ptr** to access the LEDs that you would otherwise have no way of accessing.

A further example should help to clarify the use of pointers in C.

```
/* pointer.c */
#include <stdio.h>
#include <conio.h>

int main(void)
{
    int data = 123; /* an integer variable */
    int *ptr;       /* a pointer variable */
                    /* (points to an integer) */
    ptr = &data;    /* get the address of data */
    printf("data contains %d\n", data);
    printf("data contains %d\n", *ptr);
    printf("Address of data is %d\n", &data);
    printf("Address of data is %d\n", ptr);
    printf("Address of ptr is %d\n", &ptr);

    getch();
    return 0;
}
```

The contents of the variable **data** are first printed out in the manner that you are used to. The second **printf()** statement in the program also prints out the value of **data**, but this time by using a pointer. The address of the variable **data** was saved to the pointer variable **ptr**. We are dereferencing the pointer when we use ***ptr** in the **printf()** statement. ***ptr** means “Get the data that the address stored in **ptr** is pointing to”.

The third **printf()** statement prints out the address of the variable **data**. After this the contents of the variable **ptr** is printed which is also the address of the variable **data** as we would expect because we assigned it this value at the beginning of the program. In other words **ptr** is holding the address of the variable **data**. Finally the address of the pointer variable **ptr** is printed out. **ptr** is also a variable so it must be located in memory. If it is located in memory, it must have an address. This pointer variable just contains the address of an integer. See **figure 7.7** for a graphical explanation of the `pointer.c` program.

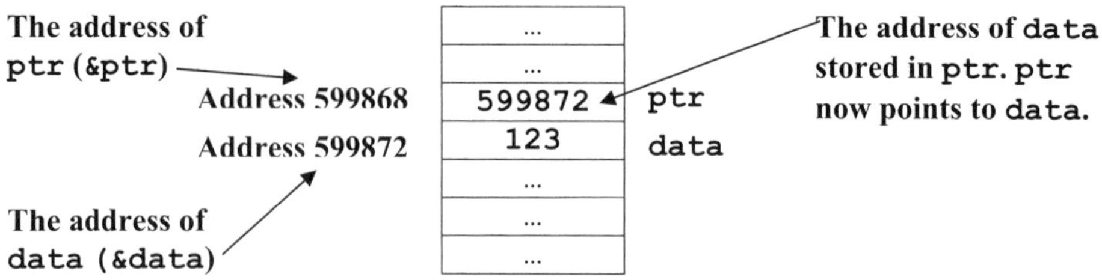

Figure 7.7 The `pointer.c` Programs Pointer and Integer Variables in Memory

Note in **figure 7.7** that the memory locations that are shown are 32 bits wide – in other words 4 bytes wide, as memory is always referred to in bytes. The pointer variable **`ptr`** and the integer variable **`data`** are shown consecutively in memory. This will not necessarily always be the case as the linker will decide where to locate these variables in memory. Because these variables are placed consecutively in memory in this example and are each 32 bits wide, the addresses are separated by 4 bytes (or 4 addresses), this is why the second address is 599872 (599868 + 4) and not 599869 (599868 + 1). Getting back to the program counter in a microprocessor, in an 8 bit processor the address in this register will be incremented by 1 each time to fetch each 8 bit instruction from memory. A 32 bit processor will fetch 32-bit wide instructions, but addressing is still in bytes so the program counter will be incremented by 4 each time to get the next 32-bit instruction (4 bytes).

Hardware on an embedded system consists of registers. A register is just like a single memory location on a processor. These registers are accessible in a C program by using pointers. Hardware is at a fixed address and we therefore can't use a variable name to access the hardware, but must use a pointer that contains the address of the hardware and thereby points to it. This will become clearer in the chapters that follow.

7.5 More on C Data Types

So far we have seen three data types: int, float and char. Each of these data types have a fixed width and can also be used in conjunction with other keywords to change whether they can be positive and negative (signed) or are positive only numbers (unsigned). Their size determines how many bytes of memory they will occupy and their sign determines what range of values they can hold.

The size of an integer is not fixed in the C standard, but depends on how the compiler defines it. The integer size is usually the same size as the maximum data size of the processor that it runs on. A 16 bit computer will usually have a 16 bit integer size; a 32 bit computer will usually have a 32 bit integer size. An integer is by default a signed integer i.e. it can store positive or negative numbers.

Memory on a 32 bit computer is 32 bits wide. This means that four bytes can be accessed with one read or write. An 8 bit computer can only read or write 8 bits at a time (one byte).

A signed 32 bit integer can store numbers in the range of -2,147,483,648 to + 2,147,483,647. $(-(2^{32} \div 2)$ to $(2^{32} \div 2) - 1)$. An unsigned 32 bit integer can hold numbers from 0 to 4,294,967,295 $(2^{32} - 1)$. This is because a single bit is used as the sign bit and is not needed when the number is unsigned.

Even a char data type holds an integer value as we have already seen. The integer value can be displayed as a character if desired and character values can be assigned to it, but they are all ultimately integer numbers on the computer. **Table 7.1** shows the integer data types in C. The sizes in the table are the actual sizes of these variables in C programs compiled with the 32 bit GNU C compiler used on the PC and the GNU C compiler used with ARM microcontrollers.

Type	Size	Min. value	Max. value
`char`	1 byte	-128	127
`unsigned char`	1 byte	0	255
`short`	2 bytes	-32,768	32,767
`unsigned short`	2 bytes	0	65,535
`int`	4 bytes	-2,147,483,648	2,147,483,647
`unsigned int`	4 bytes	0	4,294,967,295
`long`	4 bytes	-2,147,483,648	2,147,483,647
`unsigned long`	4 bytes	0	4,294,967,295

Table 7.1: C Integer Data Types

The fact that an **`int`** data type and a **`long`** data type are the same length may seem odd, but this is dependent on the compiler and the processor architecture on which the program is run, some compilers may use a 16 bit **`int`** and 32 bit **`long`**. The keyword **`signed`** can also be used with each data types, but the compiler treats each data type as **`signed`** unless **`unsigned`** is specified. The **`varsize.c`** program uses the **`sizeof`** keyword to obtain the size of each data type in bytes. You can use this keyword if you are not sure what size the data types are on the computer/compiler that you are using.

```
/* varsize.c */
#include <stdio.h>
#include <conio.h>

int main(void)
```

```
{
    printf("On this computer,\n");
    printf("A char is %lu byte, thats %lu bits\n",
                    sizeof(char), sizeof(char) * 8);
    printf("A short is %lu bytes thats %lu bits\n",
                    sizeof(short), sizeof(short) * 8);
    printf("A int is %lu bytes thats %lu bits\n",
                    sizeof(int), sizeof(int) * 8);
    printf("A long is %lu bytes thats %lu bits\n",
                    sizeof(long), sizeof(long) * 8);

    getch();
    return 0;
}
```

This program prints out the size of each data type in bytes and in bits. The **`printf()`** statement is split over two lines so that it fits better on the page. This is fine in C as the extra spaces or tabs are ignored by the compiler. The format specifier used is **`%lu`** which stands for long unsigned.

The **`vardef.c`** program shows how to use each of the integer types to display a number.

```
/* vardef.c */
#include <stdio.h>
#include <conio.h>

int main(void)
{
    char            ch = -100;
    unsigned char   uch = 0xff;
    short           shrt = 20;
    unsigned short  ushrt = 0xffff;
    int             in = -10000;
    unsigned int    uin = 0xffffffff;
    long            lng = 100000001;
    unsigned long   ulng = 0xffffffff;

    printf("char           = %d\n", ch);
    printf("unsigned char  = %d\n", uch);
    printf("short          = %d\n", shrt);
    printf("unsined short  = %d\n", ushrt);
    printf("int            = %d\n", in);
    printf("unsigned int   = %u\n", uin);
    printf("long           = %ld\n", lng);
    printf("unsigned long  = %lu\n", ulng);

    getch();
    return 0;
}
```

This program shows an example of each of the integer data types. The unsigned values are initialised with the maximum number of binary 1s that each type can hold. When the value is printed out, it can be seen that it corresponds to the maximum value shown in **table 7.1**. The format specifier **`%ld`** (long decimal) is used to print out the long data type and the **%lu** (long unsigned) format specifier is used to print out the unsigned long data type. The **l** in these format specifiers is known as a modifier as it modifies the decimal and unsigned data types to long.

Figure 7.8 shows what the data types of the above program look like in memory. Each of the cells in the diagram represents one byte, so an **int** will take up four of these cells to make up its 32 bits. If you look at **figure 7.6** and **figure 7.7**, you will notice that each cell in these diagrams represented 32 bits. Note that the addresses in **figure 7.8** and the other examples could be any address in memory that the computer loads these variables to when the program is run and is not necessarily the exact addresses displayed here. However the bytes that make up a number that is larger than 8 bits will always be successive bytes in memory e.g. if the first byte of a 32 bit number occurs at address 1000, the other 3 bytes will occur at the addresses 1001, 1002 and 1003 to make up all 32 bits of the number.

<table>
<tr><th>Address</th><th>Data in Memory</th><th>Data Type</th></tr>
<tr><td>50188</td><td>-100</td><td>char ch</td></tr>
<tr><td>50189</td><td>0xff</td><td>unsigned char uch</td></tr>
<tr><td>50190</td><td rowspan="2">20</td><td rowspan="2">short shrt</td></tr>
<tr><td>50191</td></tr>
<tr><td>50192</td><td rowspan="2">0xffff</td><td rowspan="2">unsigned short ushrt</td></tr>
<tr><td>50193</td></tr>
<tr><td>50194</td><td rowspan="4">-10000</td><td rowspan="4">int in</td></tr>
<tr><td>50195</td></tr>
<tr><td>50196</td></tr>
<tr><td>50197</td></tr>
<tr><td>50198</td><td rowspan="4">0xffffffff</td><td rowspan="4">unsigned int uin</td></tr>
<tr><td>50199</td></tr>
<tr><td>50200</td></tr>
<tr><td>50201</td></tr>
<tr><td>50202</td><td rowspan="4">100000001</td><td rowspan="4">long lng</td></tr>
<tr><td>50203</td></tr>
<tr><td>50204</td></tr>
<tr><td>50205</td></tr>
<tr><td>50206</td><td rowspan="4">0xffffffff</td><td rowspan="4">unsigned long ulng</td></tr>
<tr><td>50207</td></tr>
<tr><td>50208</td></tr>
<tr><td>50209</td></tr>
</table>

Figure 7.8: Integer Data Types in Memory

There are other floating point data types that we will look at on our embedded system.

7.6 Choosing a Microcontroller and Embedded System

The introduction to this book already provided some information on what to look for when choosing an embedded system to learn C programming on. The microcontroller of choice is the AT91SAM7S256 or a member of this family of microcontrollers. To learn more about this microcontroller, we need to take a look at the microcontroller's datasheet.

The datasheet can be found on the Atmel website **www.atmel.com.** On the main page, click **PRODUCTS**, under **Microcontrollers** click **AT91SAM 32-bit ARM-based Microcontrollers**, now click **Datasheets** and then on **AT91SAM7S Series**. Download the **AT91SAM7S Series Preliminary Summary** datasheet and open it (**6175s.pdf**).

In this datasheet, a comparison of the microcontrollers from the AT91SAM7S series is given in Table 1-1 on page 3. The table is shown below (**table 7.2**):

Table 1-1. Configuration Summary

Device	Flash	TWI	Flash Organization	SRAM	USB Device Port	USART	External Interrupt Source	PDC Channels	TC Channels	I/O 5V Tolerant	I/O Lines	Package
AT91SAM7S512	512 Kbytes	Master	dual plane	64 Kbytes	1	2(1)(2)	2	11	3	Yes	32	LQFP/ QFN 64
AT91SAM7S256	256 Kbytes	Master	single plane	64 Kbytes	1	2(1)(2)	2	11	3	Yes	32	LQFP/ QFN 64
AT91SAM7S128	128 Kbytes	Master	single plane	32 Kbytes	1	2(1)(2)	2	11	3	Yes	32	LQFP/ QFN 64
AT91SAM7S64	64 Kbytes	Master	single plane	16 Kbytes	1	2(2)	2	11	3	Yes	32	LQFP/ QFN 64
AT91SAM7S321	32 Kbytes	Master	single plane	8 Kbytes	1	2(2)	2	11	3	Yes	32	LQFP/ QFN 64
AT91SAM7S32	32 Kbytes	Master	single plane	8 Kbytes	not present	1	1	9	3(3)	Yes	21	LQFP/ QFN 48
AT91SAM7S161	16 Kbytes	Master/ Slave	single plane	4 Kbytes	1	2(2)	2	11	3	No	32	LQFP
AT91SAM7S16	16 Kbytes	Master/ Slave	single plane	4 Kbytes	not present	1	1	9	3(3)	No	21	LQFP/ QFN 48

Table 7.2: AT91SAM7S Microcontroller Family Devices

I will be using the AT91SAM7S256. It is also possible to use the next microcontroller up from this which is the AT91SAM7S512. This microcontroller is identical to the '256 but has 512k bytes of Flash memory instead of the 256k bytes of Flash memory.

Not all of the microcontrollers from this family have a USB port. The USB port is recommended as it can be used to load code into the microcontroller. There are two other ways to load code into this family of microcontrollers, namely, via the DBGU serial port and via the JTAG port (which requires an external piece of hardware to do this – the JTAG ICE).

The AT91SAM7S64 as well as the AT91SAM7S128 could also be used with this book. These microcontrollers have less Flash and SRAM memory than the '256 as shown in the table, but otherwise are identical. The only problem when using these microcontrollers (the '64 and '128) with this book is that most of the example programs are set up to run from SRAM to make programming quicker and easier while learning and these micros will not have enough SRAM to run some of the examples. The example programs would have to be loaded to Flash memory instead.

You may want to read through the rest of the summary datasheet to familiarise yourself with the AT91SAM7S family of microcontrollers. If you have never read a microcontroller datasheet before, there may be a lot of things that you don't understand. Read through it anyway and see what you can pick up. For those who have already used a microcontroller before, the datasheet should make sense to you. A lot of what this datasheet covers has to do with hardware and even the kind of packaging that the microcontroller uses, so just skip over what you feel is unnecessary.

We now need to choose an embedded system that contains one of the above mentioned microcontrollers. Our minimum requirements from this book's introduction are:

1. **Four or more LEDs**. These should be interfaced directly to the pins of the microcontroller and not via an external chip such as an I^{2}C LED driver. If an external chip such as this is used, the program used to control the LEDs will have to change. Note that a chip that just buffers the pins allowing the LEDs to be driven with a higher current is fine.

2. **Four or more switches**. As with the LEDs, these must be directly interfaced to the pins of the microcontroller.

3. **An RS-232 serial port**. This must be connected to the DBGU serial port of the microcontroller. The programming will have to change slightly to use any of the other serial ports on the microcontroller.

4. **An ADC (analogue to digital converter)**. The AT91SAM7S256 microcontroller has four dedicated pins on the chip that can only be used as inputs to the ADC that is on the chip. Make sure that your embedded system brings at least one of these pins to a connector so that you can use it.

5. **An 18.432MHz crystal**. This was not mentioned in the introduction, but is necessary in order for the AT91SAM7S256 to generate the correct frequency to use the USB port. This is connected to the two oscillator pins of the microcontroller.

I will be using the standard evaluation board from Atmel, the AT91SAM7S-EK. This board contains an AT91SAM7S256 microcontroller with the following peripherals available on the board:

1. 4 push button switches interfaced directly to microcontroller pins.
2. 1 reset push button connected to the microcontroller's reset pin.
3. 1 USB device port.
4. 1 DBGU RS-232 serial port.
5. 1 additional RS-232 serial port.
6. 1 JTAG port.

7. 4 buffered analogue inputs from the ADC.
8. 1 I²C EEPROM memory chip. This part was not on the board when I bought it, so I soldered one of my own EEPROMs onto the board.
9. 1 connector (header) connected to all of the microcontrollers I/O pins.
10. Set up to use the USB port to load code into the microcontroller i.e. it has an 18.432MHz crystal and another requirement is met – the PIOA16 pin of the microcontroller is dedicated for use with the USB pull-up resistor. More information on this is given below.

The board can be powered from the USB connector or from an external power supply. The TST pin of the microcontroller is connected to a jumper. This pin is necessary so that the on-chip boot loader can be re-loaded into Flash memory and the utility used to load code to the microcontroller can be used. Information on how this works is provided in the next chapter.

Figure 7.9: The Atmel AT91SAM7S-EK Evaluation Board

Other boards are available with this same microcontroller on them and with additional hardware peripheral devices connected to them. You will need to weigh up the cost of each board as well as the peripherals on each board to decide which one you are going to purchase.

Information on the Atmel evaluation board can be found at www.atmel.com. Click **PRODUCTS** on the main page, then click **AT91SAM 32-bit ARM-based Microcontrollers** under **Microcontrollers**, next click **Tools & Software**. Click the **Evaluation Kit** link and the **AT91SAM7S-EK** link. Under **Documents**, the **AT91SAM7S-EK Evaluation Board User Guide** link will open up a PDF file that contains information on the board as well as the schematic diagram.

Another vendor of Atmel ARM boards is Olimex – **www.olimex.com**. Click the **Development Boards** link, and then **ARM**. Take a look at the menu on the left. The SAM7-P64 and SAM7-P256 boards contain the AT91SAM7S64 and AT91SAM7S256 microcontrollers respectively, otherwise the boards are identical. Note that these boards only have two LEDs and two push button switches, but if you are an electronics hobbyist it will be no problem to add some extra LEDs and switches to the board on the board's prototyping area. When adding LEDs, check the AT91SAM7S data sheet to see how much current each pin can deliver so that you do not overload the microcontroller and blow it. Olimex provides the circuit diagram for their board, so you can just copy the already interfaced LEDs and switches to add your own ones.

To look for other available boards, just use the internet to search for AT91SAM7S.

7.7 Summary

- Two basic memory types are used on embedded systems: ROM and RAM.
- ROM is non-volatile memory and keeps its contents even after the power to it has been switched off.
- RAM is volatile memory and looses its contents when the power to it is switched off.
- Memory is arranged as consecutive bytes in a computer or embedded system.
- Each byte in memory has its own address and can be accessed by putting this address on its address bus.
- Data is written to or read from memory using the data bus.
- The control bus contains signals to move data into or out of the memory chip.
- Pointers are used in C to access memory using the address of the memory.

- Each variable in C stores data and has an address in memory.
- C integer data types can be used to create and access variables of different bit widths.
- The width of each integer variable determines the size range of the number that it can hold.

8 Your First Embedded C Program

What you will do in this chapter:

- Get an overview of the embedded programming process
- Download embedded programming tools
- Install the tools
- See how to set up an embedded project
- Write, compile, load and run your first embedded C program

8.1 How Embedded Programming Differs from PC Programming

The C programming tools that you have been using up to now will produce programs that will run only on a PC that is running Windows. To program an embedded system in C, you need to install C programming tools that will produce programs that will run on the particular microcontroller architecture that you have chosen. Writing and compiling programs on the PC that will run on your chosen embedded system is known as cross-compiling.

On a PC, an executable file is output from the compile and link process. An embedded system needs a binary (**`.bin`**) or hexadecimal (**`.hex`**) output file to be produced by the programming tools. This file will then be loaded into the memory of the embedded system and run.

When programming an embedded system, you will still write your C program in a text file and the entry point of the C program will still be a function called **`main()`**. There are, however, some additional files that you will need.

Firstly you will need a start-up file that is usually written in assembly language. The purpose of this file is to set up various things that the microcontroller must have in place before running your C code. The code from the start-up file will be the first thing that is run when the microcontroller starts up by switching on the power to the system or pressing the reset button. This code will set up the microcontroller and then run your **`main()`** function. From this point on, you will be running your C program.

Any embedded system that you buy should have the assembly language start-up file and an example program with the system, so you should not have to write this file yourself. You may have to modify the file depending on how you would like the microcontroller to be set up. E.g. you may want to start the microcontroller with a different clock speed. Be aware that if the assembly language start-up file is included with your embedded system, it may not have been written for the programming tools that you are using. A number of evaluation boards have start-up files and example programs written for an evaluation version of commercial programming tools. In this case you may have to re-write the start-up file if you are using different programming tools. This is because the format of assembly language files varies from compiler to compiler. Alternatively there may be example programs available on the Internet for your chosen system and tool-chain.

Assembly language files are text files and usually have a `.s` or `.asm` file extension. The start-up file is usually called `startup`, `crt0` or `crt` i.e. `startup.s`, `crt0.s` or `crt.s` (crt stands for C run-time).

It is best to build the microcontroller project using a make file. The make file will be used to build the assembly language file by calling the assembler program from your tool-chain. It will call the C compiler to compile your C source file(s), linker to link the object files produced from the assembly language and C source files with any library files. It will also be used to run a program that will convert the output file from the linker into a hex or bin file for loading into the embedded system.

Finally you will need a linker directive or linker command file that is used by the linker to place your program at the correct address in the microcontroller's memory. This file contains the location and size of the memory on the embedded system.

Figure 8.1 shows the complete build process for an embedded C program. "Build" refers to the entire process of producing a useable output file from source and library files.

On the PC, the executable file that was produced by the C tools could be run on the target system (i.e. the PC). In an embedded system, there must be some way of loading the binary or hexadecimal program file into the microcontroller's memory so that when the microcontroller starts up, it will start running the program.

The manufacturer of the microcontroller will provide a way of loading the program file to the microcontroller for execution. Early microcontrollers never had their own on-chip non-volatile memory. They used an external memory chip that was inserted into a special chip programmer. The chip programmer would then be used to load the executable code (bin or hex file) onto the non-volatile memory chip. Once programmed, the memory chip would be removed from the chip programmer and plugged into an I.C. socket on the embedded system board. When power was applied to the board, the microcontroller would start executing instructions from the program in the external memory chip.

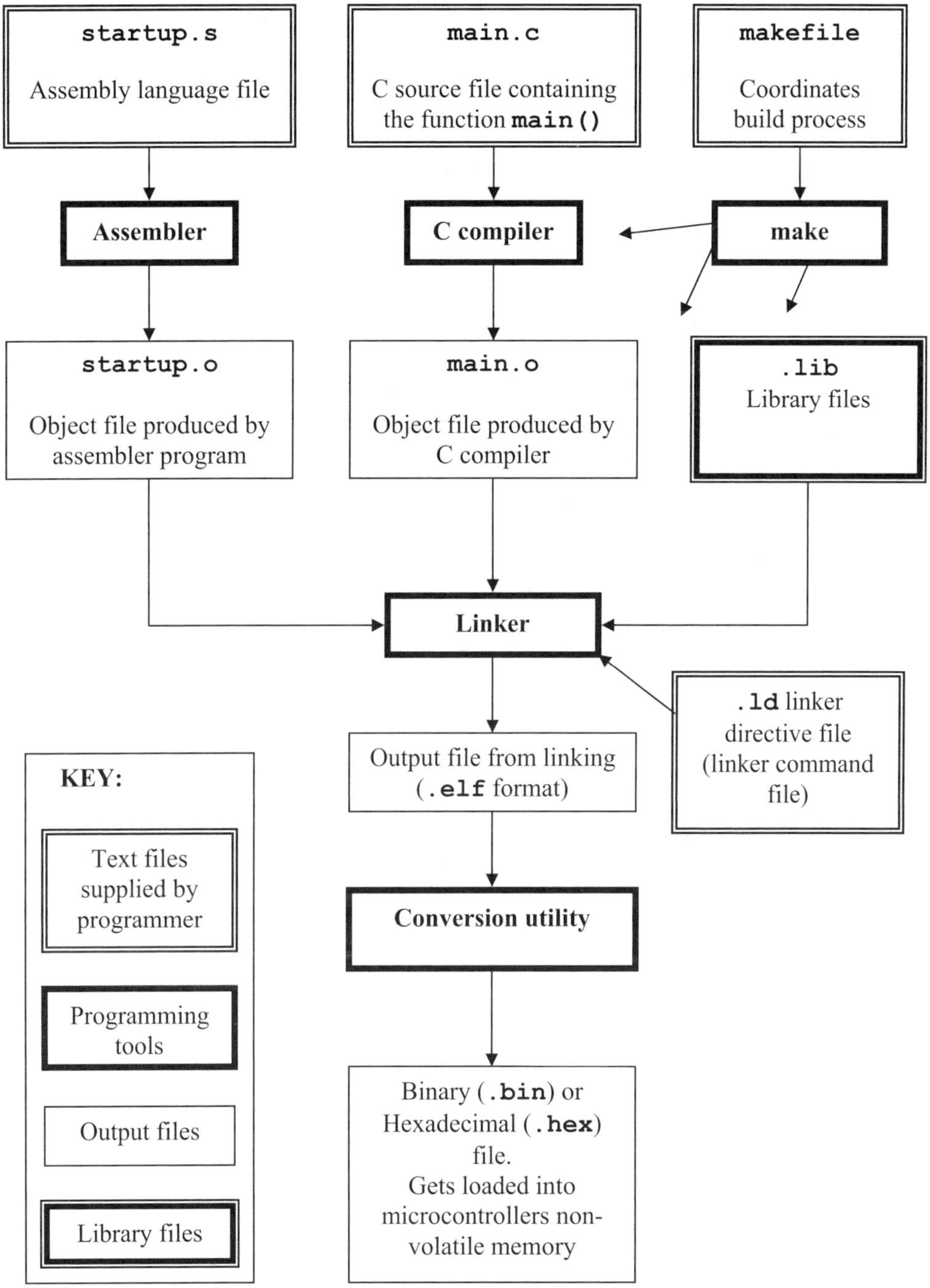

Figure 8.1: Embedded Programming Build Process

Many modern microcontrollers now have on-chip Flash memory. This is re-programmable non-volatile memory. It may be possible to still use a chip programmer, but there is usually a better option called in-system programming (ISP) for loading the executable code to the microcontroller. The Atmel AT91SAM7S256 for example has ISP capabilities. The program can be loaded onto the microcontroller via the USB port or the RS-232 serial port by using a program supplied by Atmel that runs on the PC. This is a cheap way of programming the microcontroller, as all that is needed is a USB cable or serial cable. More about this later.

8.2 The Embedded C Programming Tools

Setting up embedded programming tools will depend on the microcontroller that you have chosen. If you have bought an evaluation board or general purpose microcontroller board, it should have come with instructions and application notes. By having a pre-built and tested board you know that it is working and can get your software tool-chain up and running without the hassle of possible hardware problems.

I have chosen the ARM7 processor architecture to use with this book and the AT91SAM7S256 microcontroller from Atmel in particular. If you would like to program a different microcontroller, you will need to get the C programming tools for your particular microcontroller. These tools may or may not be free. Some programming tools are evaluation versions of commercial tools and will have some restriction on them, usually a restriction on the size of program that you can compile.

Unfortunately because the architecture of each manufacturer's microcontroller is different from each other, it is not possible to talk in general terms when programming an embedded system. Even if two different microcontrollers are ARM7 based and have LEDs interfaced to them, they may have a different way to switch the LEDs on and off or the LEDs may be at a different address on the address bus.

The good news is that the free GNU tools that we are going to install can be used with ARM7 and ARM9 microcontrollers and there are a number of manufacturers of ARM microcontroller to choose from.

We will be installing the YAGARTO toolchain from **www.yagarto.de** for programming ARM7 based microcontrollers in Windows.

8.3 The YAGARTO Toolchain

YAGARTO stands for "Yet Another GNU ARM Toolchain". Toolchain refers to the series of applications that are used to edit and compile programs, i.e. the compiler, linker, IDE, debugger, etc. YAGARTO runs on Windows PCs and uses the very popular Eclipse IDE, so we will be using the Eclipse IDE instead of Programmer's Notepad to write and build our programs.

8.3.1 Installing YAGARTO

The installation instructions of YAGARTO have changed several times, with each new version, since I have been using it. The best way to learn how to download and install YAGARTO is to read the installation instructions on the YAGARTO website. Michael Fischer who runs and maintains the YAGARTO website and toolchain has done an excellent job of updating installation instructions for each new release of YAGARTO. Note, however, that you do not have to install all of the software right away, the installation instructions on the website include the installation of debugger software that is used in conjunction with a hardware debugger device. We will be covering debuggers in a later chapter, so you can skip this step for now.

There is also an excellent tutorial written by James P. Lynch available on the Atmel website. This tutorial contains step by step installation instructions on how to install YAGARTO. The tutorial, however, may be out of date when the YAGARTO installation instructions change, but it is still an excellent tutorial on the use of YAGARTO tools for programming embedded systems and a must read. The YAGARTO website home page has a link directly to the file:

http://www.atmel.com/dyn/resources/prod_documents/atmel_tutorial_source.zip

The basic components that need to be downloaded and installed are as follows:

1. **YAGARTO tools.**

2. **YAGARTO GNU ARM toolchain.**

3. **JAVA runtime environment.**

4. **Eclipse IDE.**

I will take you through the installation of these four components as they were installed at the time of writing. Check the YAGARTO website for any differences in installation. The software to use with the hardware debugging device can be installed later.

Links to all of the software packages above can be found on the YAGARTO home page at **www.yagarto.de**. Click the **Download** link on the left menu or scroll down to see the download links.

The first item to download and install is the **YAGARTO Tools**:

yagarto-tools-20070303-setup.exe

Just download it and run it and you will be guided through the installation process. Leave all of the default items to install selected and install to the default directory. Note that the dates in the downloaded file above may be different from the one that you download.

Next, download the **YAGARTO GNU ARM toolchain**:

yagarto-bu-2.19.1_gcc-4.4.1-c-c++_nl-1.17.0_gi-6.8.50_20090817.exe

Run this file and you will again be guided through the installation process. Leave all of the default components selected and install to the default directory. Again, the dates and version numbers in the package name may differ from the one that you have downloaded.

Finally download the **Integrated Development Environment**:
This link will open a page from the YAGARTO website with instructions on how to install the Eclipse IDE. In order for Eclipse to be able to run, you must first have the Java Runtime Environment installed on your PC. Follow the instructions and first check if Java is already installed on your PC. If it is not, click the **Java SE Downloads** link to go to the Sun website:

http://java.sun.com/javase/downloads/index.jsp

On the Sun Java webpage, scroll down to find the link marked:

Java SE Runtime Environment (JRE)
JRE 6 Update 16

This link will enable you to download the Java Runtime Environment.

jre-6u16-windows-i586.exe

Download and install it. Note that the version number may be different on the link and file that you download.

Back on the YAGARTO webpage, there is a link to the Eclipse website for downloading the Eclipse IDE, this will take you to:

http://www.eclipse.org/downloads/

On this webpage, you must download the Windows version of:

Eclipse IDE for C/C++ Developers

The name of the file that you will download is:

eclipse-cpp-galileo-win32.zip

To install Eclipse, copy the **`eclipse`** directory from the downloaded zip file to the root directory of your C drive, i.e. **`C:\`**.

8.3.2 Testing the YAGARTO Installation

To check that the YAGARTO tools were installed successfully, start a command prompt by typing **cmd** in the Windows run dialog box and enter:

```
arm-elf-gcc --version
```

You should get back copyright and version information. If you encountered any problems during installation, be sure to read the latest installation instructions on the YAGARTO website.

8.3.3 Running Eclipse for the First Time

The instructions on running Eclipse are available on the YAGARTO website. When you click the **Integrated Development Environment** link under **Downloads**, the page that opens contains installation instructions for Eclipse as well as how to run Eclipse. I will take you through running Eclipse for the first time, or you can use the instructions on the YAGARTO website. The instructions on the website go on to explain a lot more about Eclipse, but you only need to do the first few steps as explained below. Look at the website page for screenshots of Eclipse if you are unsure of what to do.

To start Eclipse, double click the **eclipse.exe** icon in the eclipse directory (**C:\eclipse**). You may want to make a shortcut to this program on your desktop or start menu for easier access to Eclipse.

When Eclipse runs for the first time, it will ask you to select a workspace. In the **Workspace Launcher** dialog box, click the "**Use this as the default and do not ask again**" check box and then click the **OK** button as shown in **figure 8.2**.

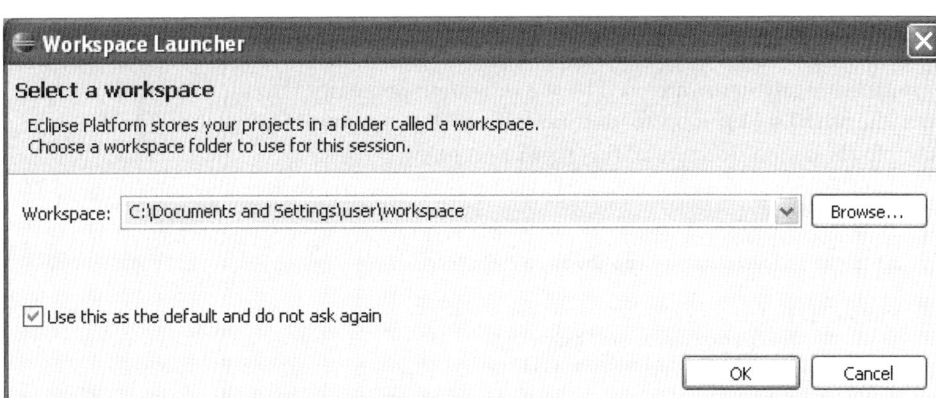

Figure 8.2: The Dialog Box that Appears when Running Eclipse for the First Time

A start up window in Eclipse will now be displayed. Click the Workbench icon on the start up window.

You are now ready to use Eclipse.

8.4 Writing Your First Embedded Program

The traditional "hello world" program used when learning a new programming language can't necessarily be written to run on an embedded system. The embedded system may not have a screen or LCD display to display the message on. We will solve this problem later by using the serial port on the embedded system and linking it to a PC that is running a terminal emulation program such as HyperTerminal on Windows. Not every embedded system will have a serial port, that is why it was recommended to buy an embedded system that does have one. Still, programming the serial port is not the simplest program to start with.

A traditional first embedded system program is to flash an LED. This assumes that the embedded system has at least one LED that can be controlled by an embedded program.

8.4.1 The `while(1)` Loop

When an embedded system program has finished running, unlike the PC, it has no operating system to return to. You will typically want to have the program running continually and never exit. If you were using a calculator for example, you would not want the calculator program to finish and then stop. You would want it to be ready for you to enter a new calculation each time.

The simplest way of stopping the program from terminating is to write the body of the program in a **`while(1)`** loop. This is just a **`while`** loop that always evaluates to true because it has 1 as its loop expression. The typical simple embedded program will start by initialising any hardware used by the program upon entry into the C code. It will then enter the **`while(1)`** loop to run the bulk of the program continually. The following example shows the structure of this type of program:

```
int main(void)
{
    /* initialise hardware here */

    while (1) {
        /* the program runs here */
        /* this loop is never exited */
    }
}
```

8.4.2 About the Source Code

The source code supplied with this book contains the example program that we are going to use. It is specifically set up for the AT91SAM7S microcontroller family and includes all the necessary files. If you have chosen a different microcontroller, it is up to you to read the documentation for the system that you have bought and find a simple example program to start with. Unfortunately I can't include code for every available microcontroller so have to leave it at that.

The example program enables you to concentrate on C programming without having to set up the start-up file, linker directive file and make file. The C source file contains an already written C program. You may need to modify it to run on your particular board.

8.4.3 Opening the Program in Eclipse

The embedded programs used with this book are available from the Elektor website at:
www.elektor.com

Refer to appendix B for details on each program used in this book.

Create a directory for storing your embedded programs in. Copy the example code directory to your embedded programs directory. The example program is called **`LED_onoff`**. I am going to create a directory called **`ARM_C`** on the D drive of my computer and copy the **`LED_onoff`** program directory into it i.e. **`D:\ARM_C\LED_onoff`**.

The **`LED_onoff`** directory contains four files:

1. **`startup.s`** – an assembly language file that initialises the microcontroller and calls the C **`main()`** function.
2. **`main.c`** – contains the C source code and **`main()`** function. I am in the habit of calling the C source file that contains the **`main()`** function **`main.c`** to make it easier to find **`main()`** when the project consists of a large number of source files. You could, of course, name this file anything that you would like to.
3. **`makefile`** – the file used with the **`make`** utility to build the project.
4. **`Flash.ld`** – the linker directive file that tells the linker where in the microcontroller memory to put the program code and data.

You may need to make changes to **`main.c`** and possibly **`makefile`**. First we will open the project in Eclipse and then see that we can build it to produce a **`.bin`** binary file. We will then examine what needs to be changed in the project's source files and then re-build the project if changes are necessary. After this we will download the utility program from the Atmel website that enables us to program our embedded system via the USB port of the computer. If you have problems building the project as described below, read the section on modifying the project as you may need to change some things in the make file in order to build correctly. The section on modifying the project follows the section on building it below.

To open the project in Eclipse:

1. Start Eclipse if you have closed it.

2. Click **File → New → C Project**.

3. In the **C Project** dialog box, uncheck the "**Use default location**" check box (**figure 8.3**).

4. Use the **Browse...** button to navigate to the project directory and click **OK** when selected. e.g. mine is `D:\ARM_C\LED_onoff`.

5. In the **Project name:** box, enter a name for the project. I am going to give the name of my project the same name as the directory that it resides in: `LED_onoff`.

6. In the **Project types:** box, make sure that **Empty Project** is selected under **Makefile project**.

7. In the **Toolchain:** box, make sure that **--Other Toolchain--** is selected

8. Click the **Finish** button.

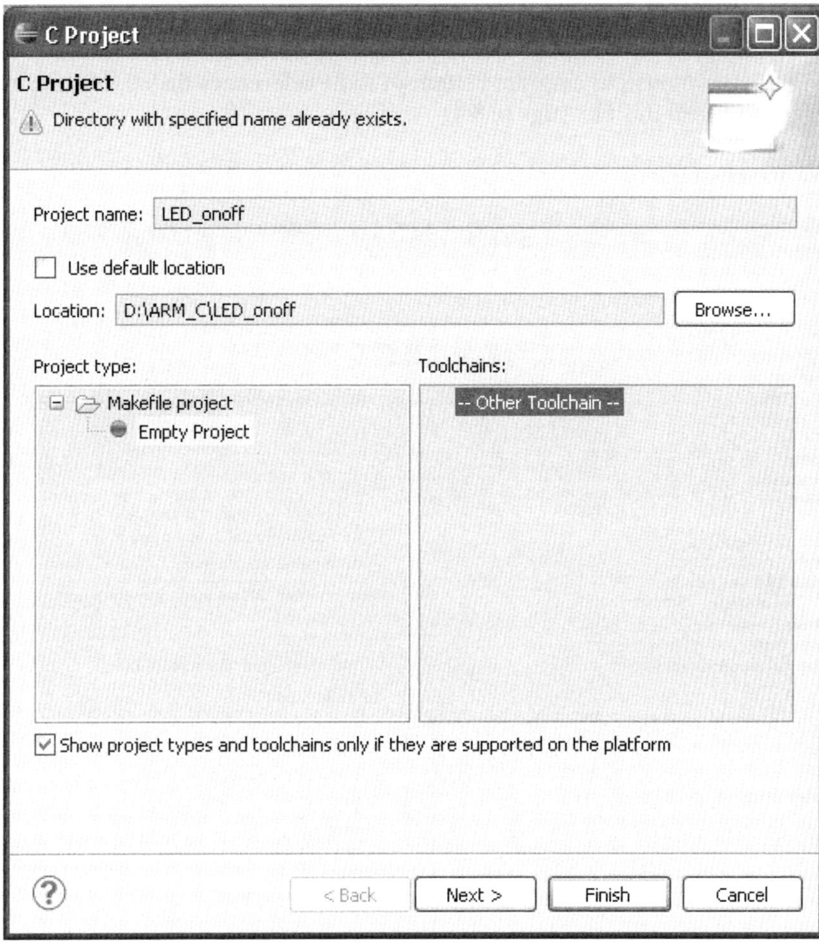

Figure 8.3: The New C Project Dialog Box

The project will be built automatically after you click the **Finish** button. A warning appears in the **Problems** pane. To fix this we need to change a setting in Eclipse:

1. Click the **Window** menu item and then **Preferences**.

2. Expand the **C/C++** item in the panel in the left of the **Preferences** dialog box by clicking the + sign to the left of it. (**Figure 8.4**)

3. Expand the "**New CDT project wizard**" item by clicking the + sign to the left of it.

4. Click on the **Makefile Project** item.

5. Click the **Discovery Options** tab at the right of the **Preferences** dialog box.

6. In the "**Compiler invocation command**" box, type **arm-elf-gcc**. If you can't see this box, use your mouse to drag the bottom of the **Preferences** dialog down to make it bigger and reveal this box (**figure 8.4**).

7. Click the **OK** button.

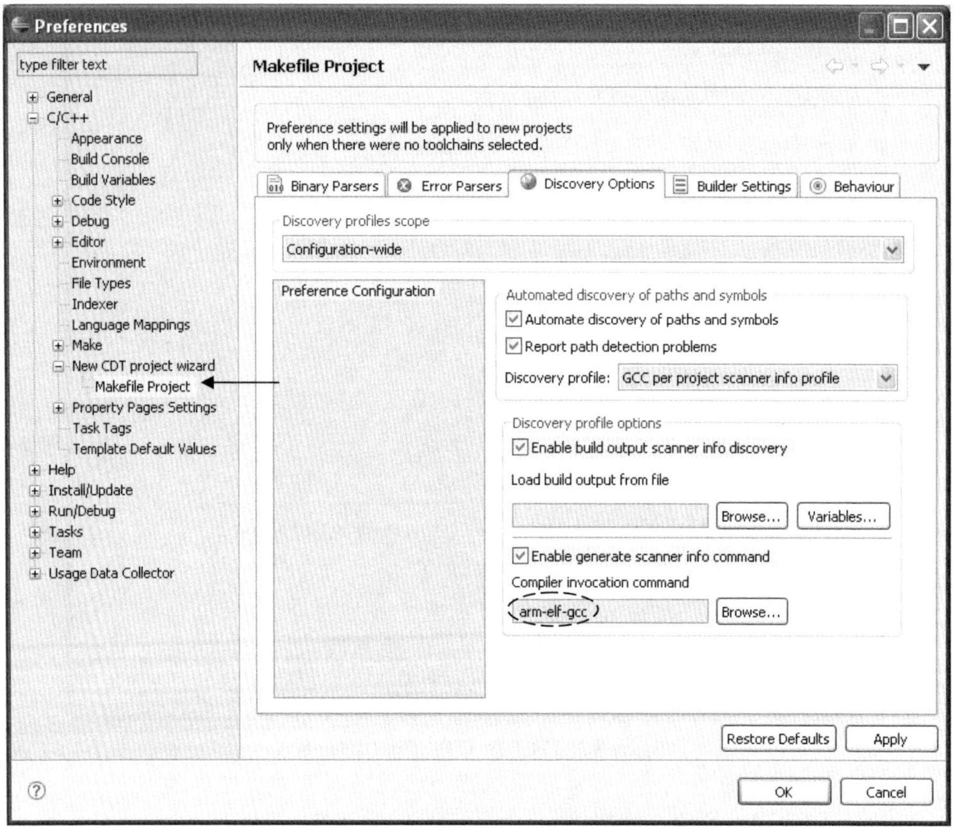

Figure 8.4: The Eclipse Preferences Dialog Box

The above steps will ensure that any new projects will not have the warning appear, but we must change this setting so that it applies to the current project as well (only for the older Eclipse version - Eclipse Ganymede):

1. In the **Project Explorer** pane in Eclipse, make sure that the project that we just created is selected i.e. **LED_onoff** in this pane must be highlighted (**figure 8.5**).

2. Now click the **File** menu and then **Properites**.

3. Expand the **C/C++ Build** item in the left pane of the **Properties for LED_onoff** dialog box by clicking the + sign to the left of it.

4. Click the **Discovery Options** item in the left pane under **C/C++ Build**.

5. Follow steps 6 and 7 of the previous setup to apply this configuration to the current project.

6. Click the **Build All** button on the Eclipse toolbar (**figure 8.5**). The warning in the **Problems** pane should now disappear.

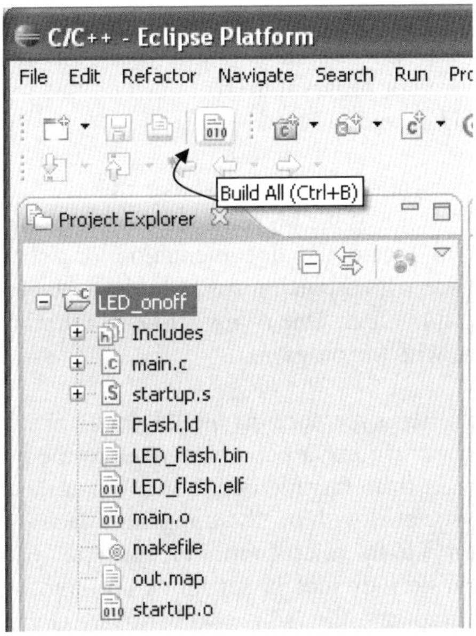

Figure 8.5: The Build All Button in Eclipse

8.4.4 Modifying the Program to Run on Your Embedded System

If you are using the same hardware as me, namely the Atmel AT91SAM7S-EK evaluation board, you will not have to modify anything in this project except for possibly the make file and then only if the gcc compiler version that you are using is different from mine (you may have downloaded a newer version of the compiler). Your embedded system must have an AT91SAM7S microcontroller on it and it must have at least one LED to be able to run this program. An embedded system with an AT91SAM7S512, AT91SAM7S256, AT91SAM7S128 or AT91SAM7S64 will be fine. If you are using one of the above microcontrollers that has less memory than the AT91SAM7S256, you will need to modify the linker directive file – this information is available right at the end of this chapter, so be sure to read this whole chapter if you are using one of these microcontrollers, before you try to run the program. If you don't modify the linker directive file for the smaller memory size, the program will not work.

If you are using a different AT91SAM7S embedded system and your LED happens to be interfaced to the same pin as the LED that I am using on my system, you will not have to modify anything in the C source file.

If you have a different ARM microcontroller to the AT91SAM7S e.g. one from NXP, you will be able to compile your C programs using the same tools that we have set up, i.e. YAGARTO and Eclipse, but you will need a different assembly language start-up file, make file and linker directive file. Your C source code file will have to use different addressing and setup different hardware to flash an LED on the different microcontroller.

The source code file must be modified to control an LED on your system. The reason that you may have to modify the source code is that if your LED is interfaced to a different pin on the microcontroller, your program needs to switch this pin high (logic 1) and low (logic 0) to switch the LED on and off.

The AT91SAM7S microcontrollers mentioned above all have 32 I/O pins labelled PA0 to PA31. PA means "port A" some of the bigger Atmel microcontrollers have a second port labelled PB. You need to locate the documentation or schematic diagram of your microcontroller board to see which pins the LEDs are interfaced to. Select one of these LEDs and note the port pin number. One of my LEDs is interfaced to port pin PA0, so I will select this LED to use with the program.

The port in the microcontroller is memory mapped. It looks like a 32 bit memory location to the microcontroller, so we can use a pointer to write to the port if we know what it's address is, which we can get from the microcontroller's datasheet[1]. The port actually has several 32 bit registers associated with it. These registers each look like a 32 bit memory location (4 bytes of memory) to the microcontroller. Each pin that is connected to the port of the microcontroller can be individually set up as an input or an output or it can be connected to one of the microcontroller's internal hardware devices. These options are all configurable through various memory mapped registers.

When the microcontroller is powered up, the port pins are set up as input pins by default. We need to write to one register to set the pin up as an output, another register to switch a pin to logic 1 so that our LED that is attached to it switches on and another register to switch the pin to logic 0 to switch our LED off.

The bits in each register are associated with the pins of the microcontroller's port as shown in **figure 8.6**. These registers can be viewed as one 32 bit long word. The figure shows one of these registers split in two halves, but this is only to fit it on the page. Two LEDs are shown interfaced to the pins of the microcontroller, pins PA0 and PA10. The LEDs can be thought of as being interfaced to bits 0 and 10 in the registers that are

---

[1] **AT91SAM7S Series Preliminary** datasheet **6175.pdf**, rev. **6175I–ATARM–24-Dec-08**. Refer to exercise 4 at the end of this chapter for download instructions. This datasheet will be referred to throughout this book.

associated with controlling these pins as outputs. To the C programmer, these LEDs look like they are interfaced to bits in a set of 32 bit memory locations. Note that a 0 or a 1 could be used to switch an LED on, depending on how they are wired. The figure shows two ways of wiring the LEDs. The LED wired to bit 0 requires a logic 0 to switch it on, the LED wired to bit 10 requires a logic 1 to switch it on.

Port Pin	PA 31	PA 30	PA 29	PA 28	PA 27	PA 26	PA 25	PA 24	PA 23	PA 22	PA 21	PA 20	PA 19	PA 18	PA 17	PA 16
Bit in register	31	30	29	28	27	26	25	24	23	22	21	20	19	18	17	16

Port Pin	PA 15	PA 14	PA 13	PA 12	PA 11	PA 10	PA 9	PA 8	PA 7	PA 6	PA 5	PA 4	PA 3	PA 2	PA 1	PA 0
Bit in register	15	14	13	12	11	10	9	8	7	6	5	4	3	2	1	0

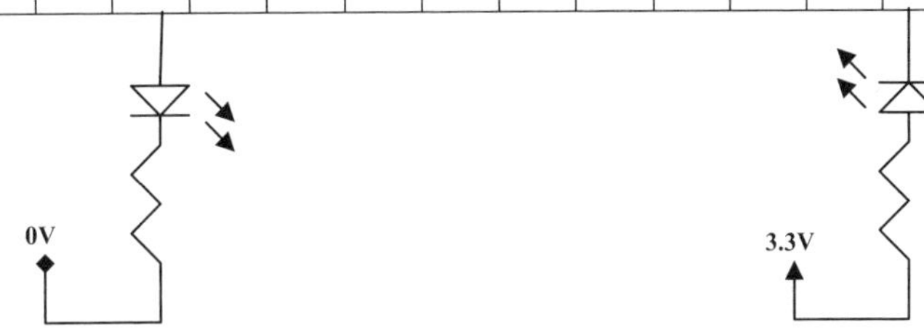

Figure 8.6: PA Port Registers on the Atmel AT91SAM7S Microcontrollers

We will now look at the C source code used to control an LED interfaced to the microcontroller. Open **main.c** by double clicking it in the **Project Explorer** pane in Eclipse. The contents of **main.c** are shown below:

```
/* AT91SAM7S256 registers used */
#define PIO_PER         (*((volatile unsigned int *) 0xFFFFF400))
#define PIO_OER         (*((volatile unsigned int *) 0xFFFFF410))
#define PIO_SODR        (*((volatile unsigned int *) 0xFFFFF430))
#define PIO_CODR        (*((volatile unsigned int *) 0xFFFFF434))

void Delay(void);

int main(void)
{
    /* configure PIO line for LED */
    PIO_PER = 0x00000001;       /* enable uC pin as PIO */
    PIO_OER = 0x00000001;       /* enable uC pin as output */

    while (1) {
            PIO_CODR = 0x00000001;
            Delay();
```

```
            PIO_SODR = 0x00000001;
            Delay();
        }
}

void Delay(void)
{
    int count = 1000000;

    /* waste time */
    while (count) {
        count = count - 1;
    }
}
```

Four pointers are defined at the top of the source file using the **#define** directive. These pointers point to hardware registers of the port controller of the microcontroller. The addresses of these registers are the hexadecimal numbers that you see in the source file. These addresses were obtained from the AT91SAM7S datasheet. The pointers are pointing to unsigned integer registers (32 bit registers) that are memory mapped on the microcontroller. The C **volatile** keyword is used with pointers to tell the compiler not to try to do any optimisation when the pointer is used – if it did, your program would not work when using the pointer.

The pointers are defined in this form:

```
(*((volatile unsigned int *) 0xFFFFF400))
```

This accesses the 32 bit data at address 0xFFFFF400 in memory by casting it to a pointer (forcing it to be a pointer to the address) and then dereferencing the pointer. This will be explained in greater detail when we revisit pointers later in the book.

In the body of the **main()** function, two registers in the microcontroller are written to:

```
PIO_PER = 0x00000001;        /* enable uC pin as PIO */
PIO_OER = 0x00000001;        /* enable uC pin as output */
```

These lines write 1 to address 0xFFFFF400 and 1 to address 0xFFFFF410. They effectively are writing 1 to the PIO Enable Register (PIO_PER) and 1 to the PIO Output Enable Register (PIO_OER). The PIO is the Parallel Input/Output Controller, a hardware device that is present on the AT91SAM7S microcontroller. This device enables the pins PA0 to PA31 to individually be used as input pins, output pins or to connect them to an embedded peripheral that is inside the microcontroller.

Writing 1 (0x00000001) to bit 0 of the PIO_PER register sets the PIO up to use pin PA0 as an input or output pin as opposed to connecting it to an internal peripheral. **Figure 8.6** shows that setting the LSB of the register (bit0) deals with configuring pin PA0 of the microcontroller.

Writing 1 to bit 0 of the PIO_OER register sets pin PA0 up as an output pin as opposed to an input pin. We will now be able to use this pin to write to the LED that is attached to it. After setting up pin PA0 of the microcontroller, the program enters the **`while(1)`** loop. Inside this loop, two more registers belonging to the PIO controller are written to. One register is used to set the PA0 pin to a logic 1 level and the other to set the pin to a logic 0 level. This will result in the LED attached to the PA0 pin switching on and off. The PIO_CODR (PIO Clear Output Data Register) clears the PA0 pin when 1 is written to it (clear = clears the pin to logic 0). The PIO_SODR (PIO Set Output Data Register) sets the PA0 pin when a 1 is written to it (sets = sets the pin to logic 1). This is how the LED is switched on and off.

A delay function is called between switching the LED on and off so that the LED remains in the on or off state for a period of time that is visible to a person using the embedded system. You have used the same delay routine in one of the PC C programs. The statements inside the **`while(1)`** loop will continually be run until the power to the board is switched off or a new program is loaded to it.

Now that you understand how the program works, you will be able to modify it to work on your AT91SAM7S embedded system. Choose which LED you would like to switch on and off and note which pin of the microcontroller it is attached to. I will use the example of **figure 8.6** where the second LED is attached to pin PA10. To switch this LED, we need to access the bit in the 32 bit wide PIO registers of the microcontroller. Refer to **figure 8.6** and write down a 0 for every pin that you do not want to use and a 1 for the pin that the LED is attached to i.e. pin PA10 in this example:

```
0000000000000000 0000010000000000
```

Now convert this binary number to hexadecimal. You can type it into your Windows Calculator. First put the calculator into binary mode and then type the number in. You do not have to type all the preceding 0s, start with the first 1 that you see when reading the number from left to right and then type in all the 0s after this i.e. type:

```
10000000000
```

Now switch the calculator into hex mode to convert the number. The alternative is to break the number into groups of four digits each starting from the right and convert each group using 8421 as follows:

```
0000 0000 0000 0000 0000 0100 0000 0000
8421 8421 8421 8421 8421 8421 8421 8421
0    0    0    0    0    4    0    0
= 0x00000400
```

The number can also be written without the preceding 0s as 0x400 in the C code.

Now all you need to do is modify the lines of code in the source file to set up and switch PA10 instead of PA0:

```
int main(void)
{
    /* configure PIO line for LED */
    PIO_PER = 0x00000400;         /* enable uC pin as PIO */
    PIO_OER = 0x00000400;         /* enable uC pin as output */

    while (1) {
        PIO_CODR = 0x00000400;
        Delay();
        PIO_SODR = 0x00000400;
        Delay();
    }
}
```

Modify your C source file to use your selected LED using the method described above. If you have problems building the project, you may need to modify the make file.

8.4.5 Modifying the Make File

The only reason to change anything in the make file is if you installed the YAGARTO toolchain into any directory besides the default or you have used a version of YAGARTO that uses different directories to store the library files in. The only lines that you would need to change are at the top of the make file named **`makefile`**:

```
LIBGCC=C:/"Program Files"/yagarto/lib/gcc/arm-elf/4.2.2/libgcc.a
LIBC=C:/"Program Files"/yagarto/arm-elf/lib/libc.a
LIBM=C:/"Program Files"/yagarto/arm-elf/lib/libm.a
```

All you would need to do is change these lines to reflect where the library files are located. You can use Windows to search for these files and then type the correct directory paths into the make file.

Build the project by clicking the **Build All** button in Eclipse (**figure8.5**) or use the **Build Project** icon (the one that looks like a hammer) to build the currently selected project.

8.4.6 Loading the Program to Flash Memory

In order to load the binary file, **`LED_onoff.bin`** that was produced by the build process, to the Flash memory in the microcontroller, we need to download a programming utility from Atmel.

On the Atmel website, **www.atmel.com**, click the **PRODUCTS** tab. Under **Microcontrollers**, click **AT91SAM 32-bit ARM-based Microcontrollers**. Click the **Tools & Software** item on the menu and then click the **In-System Programming** link on the webpage. Now click the **AT91 In-system Programmmer (ISP)** link. On the webpage that opens, scroll down and download the latest version of **AT91-ISP.exe** for Windows XP. The file that you download will be named **Install AT91-ISP v1.13.exe** with the version number possibly being different.

Run this file to install the Atmel ISP software and choose the default location to install it to. You can choose where you want the installation program to put shortcuts to the ISP software. You will be asked to reboot your computer when finished, do this now.

In order to use the SAM-BA utility that has been installed to load the binary file to the microcontroller, you need to load the on chip SAM-BA boot program into the microcontroller's Flash memory. The program already exists on the microcontroller chip, but it must be loaded into Flash by pulling the TST pin on the microcontroller high (to logic 1), providing power to the microcontroller for 10 seconds, removing power and then pulling the TST pin low (to logic 0). The next time that power is applied to the microcontroller, it will run the on-chip boot loader. It is now ready to connect to the SAM-BA boot utility program running on the PC.

As mentioned in chapter 7, the microcontroller board must be configured to use the SAM-BA utility to load programs via the USB port. The following things must be in place for the utility to work:

1. The crystal used with the microcontroller must be an 18.432MHz crystal in order for the boot loader to be able to run the USB port at the correct USB frequency of 48 MHz.

2. The TST pin must be connected to a jumper, so that you can pull the pin high to load the boot loader to Flash memory.

3. Pins PA0, PA1 and PA2 must be at a logical level of 1 when pulling the TST pin high.

4. Pin PIOA16 (PA16) must be dedicated to the USB Pull-up.

These are all hardware requirements and are mostly out of your control and dependent on the board layout and configuration. You could change the crystal on the board, but some of the other requirements may be difficult to modify on your board. It is also possible to load program code using SAM-BA with the DBGU serial port of the microcontroller. To summarise, the following has been extracted from the AT91SAM7S datasheet:

The SAM-BA™ Boot Recovery restores the SAM-BA Boot in the first two sectors of the on-chip Flash memory. The SAM-BA Boot recovery is performed when the TST pin and the PA0, PA1 and PA2 pins are all tied high for 10 seconds.

The SAM-BA Boot Assistant is a default Boot Program that provides an easy way to program in situ the on-chip Flash memory. The SAM-BA Boot Assistant supports serial communication through the DBGU or through the USB Device Port. (The AT91SAM7S32/16 have no USB Device Port.)

- **Communication through the DBGU supports a wide range of crystals from 3 to 20 MHz via software auto-detection.**

- **Communication through the USB Device Port is limited to an 18.432 MHz crystal. (The SAM-BA Boot provides an interface with SAM-BA Graphic User Interface (GUI).**

- **SAM-BA boot copies itself in the SRAM and uses a block of internal SRAM for variables and stacks. The remaining available sizes for the user codes are as follows: 57344 bytes for AT91SAM7S512, 57344 bytes for AT91SAM7S256, 24576 bytes for AT91SAM7S128, 8192 bytes for AT91SAM7S64, 2048 bytes for AT91SAM7S321 and AT91SAM7S32, 3840 bytes for AT91SAM7S161 and AT91SAM7S16.**

- **USB requirements: (Does not pertain to AT91SAM7S32/16)**
  – 18.432 MHz Quartz
  – PIOA16 dedicated to the USB Pull-up

I will be describing how to use the USB port to load program code to the microcontroller. To use the serial port, you will need a serial port cable and a serial port on your computer. As all new computers have USB ports, but not all have serial ports it is better to use the USB port.

After you have loaded the **`LED_onoff`** example program to Flash memory, the boot loader will be overwritten and you will need to re-load the boot loader to be able to load program code to the microcontroller again. Read the documentation that came with your embedded system to see which jumper is connected to the TST pin of the microcontroller. I will summarise the whole process again below:

1. With power to the board switched off, connect the jumper that pulls the TST pin high in place.

2. Switch the power to the board on. If your board is USB powered, plug the USB cable into your PC and the other end into your embedded system.

3. Wait at least 10 seconds.

4. Switch the power to the board off. Remove the USB cable from the board if it is USB powered.

5. Remove the TST pin jumper.

6. Switch the power to the board on again and plug in the USB cable. If the board is USB powered, this is done by plugging the USB cable into it.

7. This step will only need to be done the first time that you connect the USB cable to the board when it is running the SAM-BA boot loader. A **Found New Hardware Wizard** dialog box will appear. Click the **No, not this time** radio button and then the **Next** > button. Leave the **Install the software automatically (Recommended)** radio button checked and press the **Next** > button. A **Hardware Installation** dialog box will appear, click the **Continue Anyway** button. Click the **Finish** button.

8. Start the Atmel SAM-BA PC program that you installed. There should be a **SAM-BA v2.9** icon on your desktop or under **start** → **All Programs** → **ATMEL Corporation** → **AT91-ISP v1.13** → **SAM-BA v2.9**.

9. A dialog box will appear. Select **\usb\ARM0** in the **Select the connection:** drop down box.

10. Select **AT91SAM7Sxxx-EK** from the **Select your board:** drop down box where **xxx** is 64, 128, 256 or 512 depending if your board has the AT91SAM7S64, etc. as its microcontroller. My board has an AT91SAM7S256 microcontroller, so I will select **AT91SAM7S256-EK**.

11. Click the **Connect** button.

12. In the middle of the application's window that opens, the Flash tab is selected. Next to the **Send File Name:** box is an "open folder" icon. Click it and then navigate to the **`LED_onoff.bin`** file that the build process created. E.g. mine is in **`D:\ARM_C\LED_onoff`**.

13. Select **`LED_onoff.bin`** and click the **Open** button.

14. Now click the **Send File** button that is next to the "open folder" button that you clicked.

15. A dialog box will pop up asking you if you want to unlock involved regions. Click **Yes**.

16. Another dialog box will pop up asking you if you want to lock involved regions. Click **No**.

17. Your microcontroller has now been programmed. You can click the **Compare sent file with memory** to verify that the programming was successful. Note that if you locked involved regions in the previous step, the comparison will fail. Click **OK** to close the dialog box.

18. Now push your boards reset button or switch off the power to the board and then switch it on again. The LED that you chose should start flashing if everything was done correctly.

If you have any problems read the tutorial by James P. Lynch from the Atmel website mentioned under "Installing YAGARTO" in this chapter. Search for SAM-BA in the tutorial file. Skip over the sections where he sets it up to run from within Eclipse. There are plenty of screen shots and step by step instructions in the tutorial to help you.
If you bought the Atmel JTAG SAM-ICE device that is used for programming and debugging Atmel ARM microcontrollers, you don't have to load the SAM-BA boot utility to the microcontroller's memory. Install the drivers that came with the SAM-ICE on your PC, plug the SAM-ICE into a USB port on your PC and then plug the 20 way IDC connector on the SAM-ICE into the 20 way JTAG IDC connector on the embedded system. The SAM-BA PC utility will now find the SAM-ICE when it starts up and you can select this device to program your board instead of using the plain USB connection. This saves you from having to restore the SAM-BA boot utility to the microcontroller's Flash memory each time that you want to program it. Note that the Atmel SAM-ICE only works with Atmel ARM microcontrollers.

8.4.7 A Brief Explanation of the Project Files

`main.c`

The C source file already explained earlier in this chapter.

`startup.s`

The first part of this ARM assembly language file contains a number of lines starting with `.equ`. This stands for "equate" and is similar to the **`#define`** directive in C as it gives meaningful names to various numbers.

The part of the file starting with **`.text`** contains the actual ARM assembly language instructions that will run when the microcontroller is powered up. The first assembly instructions form part of the vector table. The first item in the vector table is the very first assembly language instruction that runs. The rest of the vector table consists of vectors for interrupts as we will see later in the section on interrupts.

The rest of the assembler instructions in the file do the following:

1. Sets up various microcontroller on-chip hardware devices.
2. Sets up the ARMs stacks for each mode – stacks are explained in James tutorial.
3. Moves the values of variables that are to be stored in RAM from ROM into RAM. Remember that the entire program is loaded into Flash by the boot utility. When the program starts up, it needs to initialise any variables that you have used in your program. These variables exist in RAM, and must be initialised by values that are initially stored in ROM.
4. Branches to the C **`main()`** function to run the C program.

`Flash.ld`

Tells the linker at which addresses in the microcontroller the Flash and SRAM memory is found as well as what size the memory is. The location and size of the memory on the microcontroller is found in the AT91SAM7S datasheet and is known as the microcontrollers "memory map". This linker directive file has been set up for the

AT91SAM7S256 microcontroller so if a microcontroller from this family with less memory is used, such as the AT91SAM7S64, it will require these parameters to be modified. In the **Flash.ld** file, change the following lines:

```
MEMORY
{
    CODE (rx) : ORIGIN = 0x00100000, LENGTH = 0x00040000
    DATA (rw) : ORIGIN = 0x00200000, LENGTH = 0x00010000
}
```

To:

```
MEMORY
{
    CODE (rx) : ORIGIN = 0x00100000, LENGTH = 0x00010000
    DATA (rw) : ORIGIN = 0x00200000, LENGTH = 0x00004000
}
```

The linker directive file will now be set up for the AT91SAM7S64 memory sizes, but will also work for any of the microcontrollers that have more memory than this one has. This is because the extra memory in the other microcontrollers is just ignored by the linker as we have not told it about this extra memory.

The address of the top of the stack must be changed in the **startup.s** file if the size of the memory in the linker directive file is changed. The following lines must be changed:

```
/* top of the stack in RAM */
    .equ    TOP_STACK,      0x00210000
```

To:

```
/* top of the stack in RAM */
    .equ    TOP_STACK,      0x00204000
```

makefile

The make file runs the assembler program to produce an object file from the assembly language file. It runs the C compiler to produce an object file from the C source file. It runs the linker to link the two object files with library files to produce an **.elf** output file. It then calls the **arm-elf-objcopy** to generate a binary file from the output file. The binary file is the file that is loaded into the microcontroller's Flash memory.

8.5 Exercises

1. Spend some time on the YAGARTO website. The website includes extra information like setting up a debugger. Read as much as you can to learn more about YAGARTO, Eclipse and ARM microcontrollers. **www.yagarto.de**

2. Read through James P. Lynches excellent tutorial. Remember that the installation instructions may differ slightly if the tutorial has not been updated to use the latest version of YAGARTO. The differences will be minor and there is still a lot of excellent information in this tutorial that is essential. Skip over what you feel is not necessary, like installation instructions if you have already done the installation. Debugger installation is also covered in the tutorial and may be of interest. The tutorial, as already mentioned in this chapter is at:
http://www.atmel.com/dyn/resources/prod_documents/atmel_tutorial_source.zip
There is also a link to it on the YAGARTO home page.

3. Read the documentation that came with your embedded system. See if you can load any pre-built binary programs to your microcontroller that came with your embedded system.

4. Download the full AT91SAM7S data sheet from Atmel. The datasheet can be found on the Atmel website **www.atmel.com**. Click the **PRODUCTS** tab and then **AT91SAM 32-bit ARM-based Microcontrollers** under **Microcontrollers,** now click **Datasheets** and then on **AT91SAM7S Series**. Download the **AT91SAM7S Series Preliminary** datasheet **(6175.pdf)**. The first few pages will look the same as the **Preliminary Summary** datasheet that you downloaded in the previous chapter. Read through chapter 15 in the datasheet. This chapter contains information on the Parallel Input/Output Controller (PIO) that we used in our **`LED_onoff`** program to flash the LED.

Don't worry about what you don't understand in the reading material above. Read through it again later and while reading the next few chapters of this book. It should slowly start to become clearer.

8.6 Summary

- A different set of C programming tools is required for each different microprocessor architecture.

- Embedded programs usually require an assembly language start up file to get the embedded system ready to run the C program.

- A linker directive file is necessary to tell the linker where the microcontroller's memory is located. This allows it to correctly locate the program in the microcontroller's memory.

- A bin or hex file is produced from building the embedded program. This file can then be loaded into the microcontroller's memory.

- A `while(1)` loop is the simplest way to stop an embedded C program from exiting. An embedded program should normally never be exited as it gives the embedded system its functionality.

- To switch an LED on and off that is connected to a microcontroller's port pin requires knowledge of the microcontroller port's hardware. The hardware information is found in the microcontroller's datasheet.

- The AT91SAM7S microcontroller family has a built in SAM-BA boot assist program that can be loaded into the microcontroller's Flash memory. When the embedded system containing the AT91SAM7S microcontroller is powered up and connected to a PC running the SAM-BA PC utility program, code can be loaded into the microcontroller's Flash memory. You will see in the next chapter that a program can also be loaded into and run from SRAM memory using the SAM-BA boot loader.

This chapter has covered a lot of material in order to get an embedded system to flash a single LED. I hope that you will now appreciate why the first chapters were used to teach some basic C programming on a PC. I hope that you will also understand why it was necessary to choose a specific microcontroller family to use with this book. If you have a different microcontroller you will have to load different programming tools to program it. Even other microcontrollers with ARM7 cores require an understanding of how to program their hardware before you can flash a single LED. This information is obtained from the microcontroller's datasheet, or from resources that came with your embedded board such as sample source code and documentation.

9 Embedded I/O & Memory Maps

What you will do in this chapter:

- Load a program to and run it from SRAM
- Work with output by writing to more than one LED
- Work with input by reading the switches on the embedded system
- See how hardware and memory is mapped in the microcontroller
- Learn more about the PIO controller

9.1 Loading a Program into SRAM

We will now use a template to create our projects. The template is a set of four files as we have used in the previous chapter, plus an extra header file. The template files contain a make file, start up assembly language file, linker directive file, C source file with a few lines of code and a header file with hardware addresses defined in it. The idea is to start each project with these files so that you don't have to re-write the same things that are going to be used in every project; I have already done this for you by providing you with the template files. The make file would need to be modified if you added more C source files to your project.

The template files have been set up for the AT91SAM7S256 microcontroller, so if you are using a microcontroller from this family that has less memory, you will need to modify memory sizes in the linker directive file and the address of the top of the stack in the start-up file as explained in the previous chapter.

From now on, we will be loading our programs into SRAM memory. The advantages are:

1. The program loads quicker to SRAM memory than to Flash memory.
2. You don't have to re-load the SAM-BA boot loader after each embedded program that you load, as the Flash is not overwritten.
3. The Flash memory can be programmed a limited number of times as specified in the datasheet. Although it can be programmed many times, your Flash will be preserved and last longer by programming the SRAM instead. Remember that every time you use the SAM-BA boot loader to load a program to the Flash memory, you are programming the Flash twice. Once to load SAM-BA into the Flash and then to load your own program into Flash and overwrite SAM-BA.

4. Programs run faster in SRAM memory. This may also be a disadvantage as you may test your program in SRAM and then load it into Flash. If it is a very time critical program, it may not work correctly when running from the slower Flash memory. This won't be a problem when running the programs in this book and may only be a problem with advanced programs. It can also be overcome by running some parts of a program in Flash memory and time critical parts of the program in SRAM memory.

When running programs from SRAM memory by using the SAM-BA program, the microcontroller is not in the same state that it will be in when it boots up and has not run any program yet. The SAM-BA boot program will set up the hardware that it uses and this hardware remains set up when we run our program in SRAM. We should actually write to all the hardware that SAM-BA has set up and put the hardware back into it's default state, but the programs that we are going to run work fine without doing this, so to keep things simple at this stage, we will leave the microcontroller in this state.

To load a program into SRAM memory, all we have to do is modify the linker directive file to tell the linker to link the entire program for SRAM and ignore the Flash memory altogether. This has already been done in the template files. Refer to appendix B for the template files to use with each chapter as well as the complete source code for each example program.

Our next embedded program follows and I will explain step by step what to do with the template files and how to use the SAM-BA utility to load the program to SRAM and run it.

9.2 Writing to More than One LED

We will first set up an empty program from the template files in Eclipse and then write our C program that uses more than one LED.

You will need to create a directory to save the five template files to. I am using a directory on my D drive called ARM_C to store my embedded programs in, so I will first create a subdirectory in ARM_C called LED_two i.e. **`D:\ARM_C\LED_two`**.

Next, copy the five template files into the new directory. **`D:\ARM_C\LED_two`** will now contain:

`main.c`
`startup.s`
`RAM.ld`
`makefile`
`at91sam7s.h`

Now create a new project in Eclipse:

1. Start Eclipse and then click **File → New → C Project**.

2. In the **C Project** dialog box that appears, uncheck the **Use default location** checkbox.

3. Click the **Browse...** button, navigate to your new project's directory e.g. **D:\ARM_C\LED_two**, and click the **OK** button.

4. In the **C Project** dialog box, type a project name in the **Project name:** box. I am going to give the project the same name as the directory in which it was created i.e. **LED_two**.

5. In the **Project types:** box, make sure that **Makefile** project is selected.

6. In the **Toolchain:** box, make sure that **--Other Toolchain--** is selected.

7. Click the **Finish** button.

If you would like to open the completed project from the zipped source files instead, follow the above steps, but instead of creating a directory and then copying the template files to it, copy the entire project directory.

The project should build automatically after it has been opened in Eclipse. To stop the project from building automatically, click the **Project** menu and then click **Build Automatically** so that the check mark next to this menu item is removed. I do this as I prefer to have control of when the project is built, you can leave the project to build automatically if you prefer.

Open the **makefile** file for editing by double clicking it in the **Project Explorer** pane in Eclipse. Change the following line near the top of the make file:
OUTFILE_SRAM=proj_name
to
OUTFILE_SRAM=LED_two
This will result in the output files that are generated when building the project having the name of the project. E.g. the files will have the names **LED_two.elf** and **LED_two.bin**. Save the make file.

Open the **main.c** file for editing by double clicking it in the **Project Explorer** pane in Eclipse. Add the following lines of code to **main.c** so that it looks like this (note that you will have to modify it to control the port pins that contain your LEDs if they are different from the ones in the program):

```
#include "at91sam7s.h"

void Delay(void);

int main(void)
{
    /* configure PIO lines for LEDs */
```

```
    PIO_PER = 0x00000003;          /* enable uC pins as PIO */
    PIO_OER = 0x00000003;          /* enable uC pins as output */

    while (1) {
        PIO_CODR = 0x00000001;
        PIO_SODR = 0x00000002;
        Delay();
        PIO_SODR = 0x00000001;
        PIO_CODR = 0x00000002;
        Delay();
    }
}

void Delay(void)
{
    int del = 5000000;

    while (del--);
}
```

The idea of this program is to write to two LEDs, first switching one on and the other off, then switching one off and the other on continually with a delay in-between switching states - very similar to our first embedded program. You will need to do some modifications to the program again to use the LEDs on your embedded system. If your LEDs are connected to port pins PA0 and PA1, you will not need to modify anything in the source code file; the program will run as it is.

To make the changes, first select which two LEDs you would like to use. Now write down which bits in the 32 bit word they fall on to get the number that you must use. I am going to use port pins PA11 and PA13 as an example. We know from this that the first four hexadecimal digits of the number (from the left) that we must write to the port will be zeros – i.e. bits PA31 down to PA16. The right four digits of the hexadecimal number contain the bits that line up with port pins PA11 and PA13. So the number will start with 0x0000???? We just need to work out what the lower four digits will be. The lower four hexadecimal digits are equivalent to 16 bits as follows:

PA15	PA14	PA13	PA12	PA11	PA10	PA9	PA8	PA7	PA6	PA5	PA4	PA3	PA2	PA1	PA0
0	0	1	0	1	0	0	0	0	0	0	0	0	0	0	0

As shown, we just fill in a 1 where the LEDs are attached to each port pin. The lower 16 bits of the number can be viewed as:

0010 1000 0000 0000

Convert to hexadecimal:

0x2800

The full 32 bit wide number to write to the port will therefore be:

0x00002800

The C code will now change as follows:

```
/* configure PIO lines for LEDs */
PIO_PER = 0x00002800;          /* enable uC pins as PIO */
```

```
PIO_OER = 0x00002800;        /* enable uC pins as output */

while (1) {
    PIO_CODR = 0x00000800;
    PIO_SODR = 0x00002000;
    Delay();
    PIO_SODR = 0x00000800;
    PIO_CODR = 0x00002000;
    Delay();
}
```

PA11 and PA13 are set up to be used as PIO pins by writing to the PIO_PER. PA11 and PA13 are then set up as output pins by writing to the PIO_OER. In the **`while (1)`** loop PA11 is cleared (to logic 0) by writing a 1 to bit 11 of the PIO_CODR and PA13 is set (to logic 1) by writing a 1 to bit 13 of the PIO_SODR. After the delay, PA11 is now set and PA13 cleared. This causes the LEDs to flash alternately.

After making changes to the source code, save it and then build the project using the **Build Project** button in Eclipse (the one that looks like a hammer or **Project → Build Project**). We now need to load the binary file **`LED_two.bin`** into the SRAM memory of the microcontroller using SAM-BA.

Firstly make sure that the SAM-BA boot loader is loaded into the Flash memory of the microcontroller. You can check this by plugging the board into the USB port of your PC and switching the power to the board on if it is not USB powered. If your previous flashing LED program starts running, you will know that SAM-BA has been overwritten by it. Another way to check is to start the SAM-BA PC program and then try to connect to your target board. If you can connect, the SAM-BA boot loader is loaded in the Flash memory of the microcontroller, if not, you need to re-load it. Follow the instructions in the previous chapter of using the TST pin of the microcontroller to load the SAM-BA boot loader to Flash memory. You only need to do this once when working from SRAM as we will never overwrite the boot loader by running our programs in SRAM.

With SAM-BA loaded into the microcontroller's Flash memory, do the following:

1. Plug the USB cable into your embedded system and PC. Power the embedded system up if it is not USB powered.

2. Start the SAM-BA application on the PC.

3. Connect to your board using the SAM-BA dialog box.

4. In the middle of the SAM-BA window are **Flash** and **SRAM** tabs. Click the **SRAM** tab to select it (**Figure 9.1, ①**).

5. Use the "open file" button next to the **Send File Name:** box (**Figure 9.1, ②**) to open the **`LED_two.bin`** file that was generated when you built the **`LED_two`** project.

6. Change the address to load the file to in the **Address:** box (**Figure 9.1,** ③) to **0x202000**.

7. Click the **Send File** button to load the binary file into the SRAM of the microcontroller (**Figure 9.1,** ④).

8. Now type: **go 0x202000** at the flashing prompt at the bottom of the SAM-BA window and then press **Enter** (**Figure 9.1,** ⑤).

Your program should now be running in the microcontroller's SRAM memory and flashing the two LEDs that you selected alternately. You can close the SAM-BA application as it will not be able to communicate with your embedded system now that your **LED_two** program is running. You can press the reset button on the embedded system board or switch the power off and then on again to re-load the SAM-BA boot utility. It has not been erased from Flash memory and will load each time that you power up the board and as long as you do not overwrite it. Just follow the above procedure for loading each of your programs to SRAM memory for the rest of the embedded programs in this book, unless they are made to run in Flash memory. Don't forget to click the **SRAM** tab when using the SAM-BA PC utility. Note that the SAM-BA PC utility may take a while to close after you have started your embedded program running and then click the close button on the PC application. Pressing the reset button on the embedded system will speed the closing of this application up. This is because the SAM-BA PC application is still trying to communicate with the SAM-BA boot utility on the embedded system, but the example program is now running in place of the SAM-BA boot program.

The next time that you load a program into SRAM memory, you don't even have to type the **go 0x202000** line; just push the up-arrow key on your keyboard and the last command that you typed at the prompt will appear on the screen. Just press the **Enter** key after this. Note that you will have to first click in the box so that the cursor is active there before this will work.

The SRAM memory in the microcontroller starts at address 0x200000, but the program has been linked to start at address 0x202000. The reason for this is that the SAM-BA boot loader uses some of the SRAM memory and it would load our program over this memory when we load it to SRAM. See the section "21.6 Hardware and Software Constraints" in the AT91SAM7S datasheet for details.

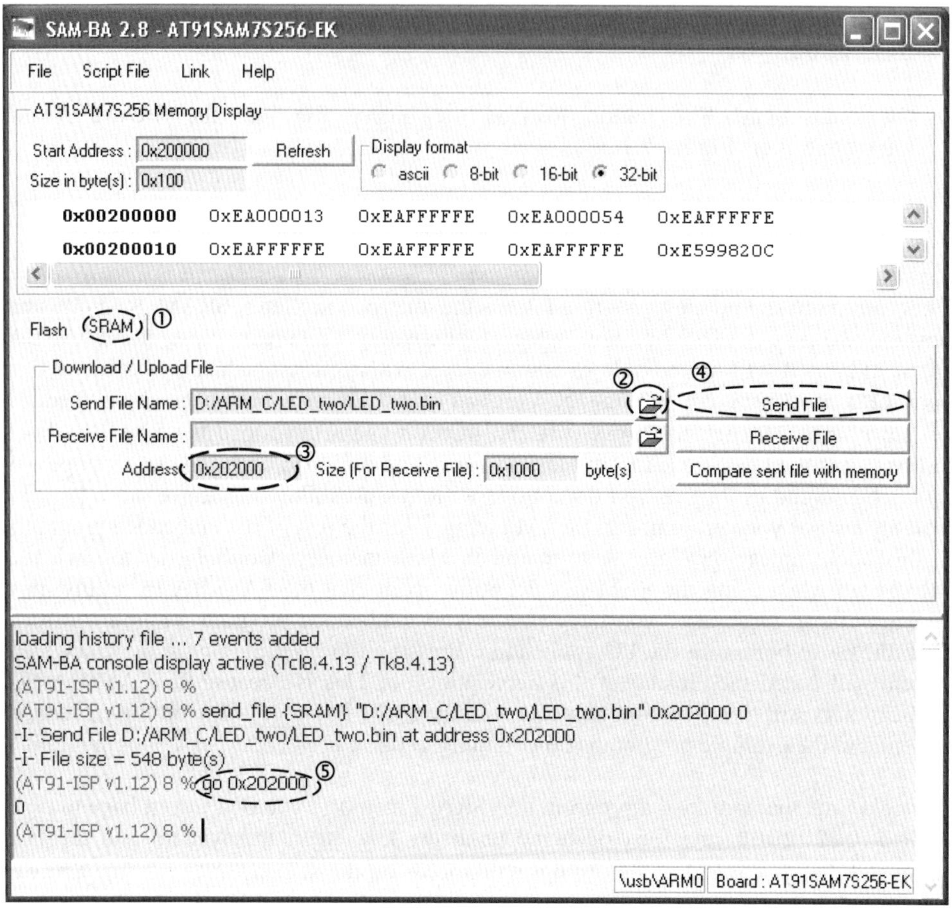

Figure 9.1: Loading the Program to SRAM Memory

9.2.1 The Program's New C Language Elements

There are some new things to you in this program that will now be explained.

The Decrement and Increment Operators

In the loop expression of the **`while()`** loop of the **`Delay()`** function we see something new:

`del--`

This is a short way of writing:

`del = del - 1`

In other words decrement (or subtract) one from the value that the variable **`del`** is holding. The `--` in C is known as the decrement operator. Placing the two minus signs after the variable name is known as post-decrementing. When the loop expression is evaluated, the value that the variable contains is first evaluated and then decremented.

It is also possible to pre-decrement the variable as follows:
--del
In this case, the value that the variable holds is first decremented by one and then evaluated when used in the loop expression. The decrement operator is not restricted to only being used in a loop expression, but can be used anywhere in a C program.

There is also an increment operator, **++**, that operates in the same way except this time adding one to the variables value.
Post incrementing a variable in C is written as follows:
del++
This is the same as:
del = del + 1
Pre-incrementing a variable in C is written as follows:
++del
and also increments the variable by one, except when being evaluated it is first incremented and then evaluated.

The **while** loop in the **Delay()** function may look a bit strange as it has no opening and closing braces:
while (del--);
This line is perfectly acceptable in C as the loop does nothing except decrement the variable **del** in order to waste time until it reaches a value of 0 which terminates the loop. Because there is no code that needs to run in the body of the loop we can write it either this way or with braces but no code between the braces.

9.3 Reading the Switches and Writing to the LEDs (I/O)

To read switches that are connected to the microcontroller's PA0 to PA31 pins, we use the microcontroller's PIO controller as we did for the LEDs. This time, however, we must set the pins used for reading switches as inputs and then read a value back from one of the PIO registers to see what the state of the switch is. The **switch_LED** program will read two switches and then switch on or off two LEDs depending on the state of the switches. Start a new SRAM project for your microcontroller in Eclipse and add the code below to **main.c**. Remember to first create a new directory e.g. **D:\ARM_C\switch_LED**, copy the correct template files to the directory and then start a new C project in Eclipse in the new directory. Open the **main.c** file for editing and add the code to it and then save the file. Also modify the make file to produce output files with the project name as we did in the previous example.

```
#include "at91sam7s.h"

int main(void)
{
    /* enable the clock of the PIO controller */
    PMC_PCER = 0x00000004;
```

```
    /* configure PIO lines for LEDs and switches */
    PIO_PER = 0x0000C003;        /* enable uC pins as PIO */
    PIO_OER = 0x00000003;        /* enable uC pins as output */
    PIO_ODR = 0x0000C000;        /* pins used as input */

    while (1) {
        if (PIO_PDSR & 0x00008000) { /* check if switch is pressed */
            PIO_SODR = 0x00000002;   /* switch out, switch LED off */
        }
        else {
            PIO_CODR = 0x00000002;   /* switch in, switch LED on */
        }

        if (PIO_PDSR & 0x00004000) {
            PIO_SODR = 0x00000001;
        }
        else {
            PIO_CODR = 0x00000001;
        }
    }
}
```

In this program a register (PMC_PCER) is used to enable the clock of the PIO controller. The PIO controller will be examined at the end of this chapter.

The PIO_PER is used to enable two extra pins as IO pins, namely PA14 and PA15. These two pins have switches attached to them on the AT91SAM7S-EK board. The other two pins set up as IO are pins PA0 and PA1 with LEDs attached to them as for the previous program.

The PIO_ODR is used to disable PA14 and PA15 as output pins. This has the effect of enabling them as input pins.

Inside the **while** loop, the two switches are continually checked to see if they have been pressed or not. If the switch attached to PA15 is pressed, the LED attached to PA1 is switched on. If the switch is not pressed, the LED on PA1 is switched off.
The same kind of **if** – **else** construct is used to check the switch on PA14 and switch the LED on PA0 on or off.

Figure 9.2 below shows how the **if** – **else** construct works in this program. It is just repeated twice in the **while** loop and checks a different switch the second time.

Is the switch closed (pushed in)?

```
if (PIO_PDSR & 0x00008000) {
    PIO_SODR = 0x00000002;  ←— No, it is not. Switch the LED off.
}
else {
    PIO_CODR = 0x00000002;  ←— Yes, it is. Switch the LED on.
}
```

Figure 9.2: The `if` – `else` Construct for Checking the State of a Switch

Before programming your hardware, you need to know how it is wired. The switches on the AT91SAM7S-EK board are wired so that they connect the IO pin of the microcontroller to ground when pushed. This means that in the un-pushed state, logic 1 will be read on the pin. When the switch is pressed, logic 0 will be read on the pin – opposite of what you would expect as 0 usually means off and 1 usually means on. This information was obtained by looking at the schematic diagram of the evaluation board.

The expression in the **`if`** statement will evaluate to true if the switch is open (not pressed) and to false if the switch is pressed.

The LEDs are wired so that when a logic level of 1 is applied to the pins that they are wired to, they will switch off. When a logic level of 0 is applied to the pins, they will switch on. This is why we set (write logic 1) to the PIO_SODR (set output data register) to switch the LED off and clear (write logic 1) to the PIO_CODR (clear output register) to switch the LED on. You will need to check the schematic diagram and/or datasheet that came with your embedded system to see how it is wired. The worst that can happen in this program is that you can get the logic mixed up and the LED will switch off when the switch is pressed on.

To modify this program for your own board, first use the same LEDs that you used in the previous program. You need to find on which microcontroller pins the switches on your embedded system are attached to. Modify the PIO_PER register contents to reflect the two LEDs and two switches that you are using i.e. in the following line of the program:

```
PIO_PER = 0x0000C003;
```

When you set up the PIO_OER register, it will be set up with the same value as in the previous program for the LEDs. Now disable the pins that you are using as inputs from the switches i.e. in this line:

```
PIO_ODR = 0x0000C000;
```

The program should work without this line as all IO pins default to inputs when the microcontroller is powered up, however it is better to include this line as a precaution.

You need to change the following line in each of the **`if`** statements in the program:

```
if (PIO_PDSR & 0x00008000)
```

The number in this line contains the bit position of the PIO pin that the switch is attached to. In the above example, bit 15 (counting from right to left with the first bit count equal to zero) is set in the number and this corresponds to port pin PA15 to which the switch is attached.

The **bitwise AND** operator is new to you here (**&**). This has nothing to do with the address operator used to get the address of a variable as you have previously seen. In this context the **&** operator is used to perform a bitwise logical operation.

You can think of the **AND** logical operation in terms of an electrical circuit as shown in **figure 9.3**.

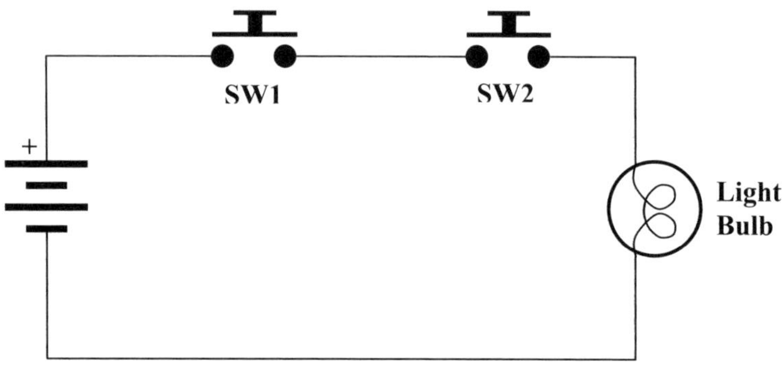

AND Truth Table		
SW1	**SW2**	**Bulb**
0	0	**0**
0	1	**0**
1	0	**0**
1	1	**1**

0 = OFF
1 = ON

Figure 9.3: The AND Circuit and Truth Table

The truth table for the circuit shows that only when switch 1 **AND** switch 2 are both on, the light bulb will switch on. All other combinations of switch 1 and switch 2 states result in the light bulb staying off. This is exactly what the C **AND** operator does. In the **switch_LED** program, the PIO register containing the state of the switch is **ANDed** with a number in order to isolate the state of the switch. Let's take an example of an 8 bit port to make things simpler. A 32 bit port will work exactly the same, except the **AND** operation is done to more bits.
If we read in a value from the port of:
`01011100`
and our switches value was bit number 3 (the fourth bit from the right) on the port, we would **AND** this with a "mask" value:
`00001000`
The result of the AND operation will be:
`00001000`

This is derived as follows:

Bit number	7	6	5	4	3	2	1	0
Port value	0	1	0	1	1	1	0	0
Mask value	0	0	0	0	1	0	0	0
Result of AND	0	0	0	0	1	0	0	0

Apply the truth table to each corresponding bit in the port value with each bit in the mask value (top to bottom). Only when we get a 1 in the port value **AND** in the mask value will we get a 1 in the result. The result tells us that the switch is on.

When placed in a C `if` statement as follows:

```
if (port_val & 0x08)
```

the result is 0x08 (`00001000B`) which is a non-zero number. The `if` statement will therefore evaluate to true and the code that is under the `if` statement will be run.

If the switch is released, the result of the C **AND** operation will now be as follows:

Bit number	7	6	5	4	3	2	1	0
Port value	0	1	0	1	0	1	0	0
Mask value	0	0	0	0	1	0	0	0
Result of AND	0	0	0	0	0	0	0	0

The result is now 0x00 (`00000000B`). The C `if` statement will now evaluate to false and the `else` statement code will run.

Note that it does not matter what is in any of the other port bits except bit 3 where our switch is attached. The other bits could all be zeros or all ones; the mask value **ANDed** with the port value will cause the result of all the other bits to be zero. The result of this **AND** operation can be shown as follows:

When the switch is on, the port value is read and then **ANDed** with 0x08, the result will be 0x08 which evaluates to true (the switch is on).

When the switch is off, the port value is read and then **ANDed** with 0x08, the result will be 0x00 which evaluates to false (the switch is off).

If there was nothing else attached to the port and the other bits always read as 0, we could get away with:

```
if (port_val)
```

The result would then always be 0x08 when the switch is on and always 0x00 when the switch is off.

We have to use the bitwise **AND** operator because there are other things on the port and we don't know what value they will be when we read the switch. In the above example, if we did not use the bitwise **AND** operator, our result of reading the port would be:
`01011100`
when the switch is on, evaluating to true and
`01010100`
when the switch is off, evaluating to true. It would then look like the switch was always on.

9.4 The AT91SAM7S Memory Map

The memory map of the AT91SAM7S microcontrollers can be found in the AT91SAM7S datasheet. The memory map shows graphically at which addresses memory and peripherals are placed as shown in **figure 9.4**.

We can see the address of each hardware peripheral on the microcontroller when we look at the memory map. Remember that each hardware peripheral that we program is just a set of registers. These registers are mapped as if they are part of the microcontroller's memory (thus memory mapped). Accessing the registers is just like accessing memory using pointers in C. We could set a pointer to write to SRAM memory and write 0xC5 to a byte of memory at this location. We could just as easily use the address of a peripheral device and write the same value to the devices register also using a pointer. The only difference is the address used.

The memory in the memory map seems to indicate that the microcontroller has 1Megabyte of Flash memory and 1 Megabyte of SRAM memory. This is not the case however; it just shows the range of addresses allocated for memory on the microcontroller. If you take the AT91SAM7S256 microcontroller, it has 256k bytes of Flash memory and 64k bytes of SRAM memory. The 256k of flash will start at address 0 of the microcontroller's memory map and fill up each address of the memory map until 256k bytes later. The SRAM will start at address 0x200000 and fill up 64k bytes of the memory map.

Just to confuse things, this microcontroller's Flash memory actually starts at address 0x100000 and SRAM memory starts at 0x200000. Either one of these memory areas can be "shadowed" at address 0. In other words it will be available at the address mentioned above and at address 0. When the microcontroller is powered up, the Flash memory is always shadowed at address 0. This is necessary as the microcontroller starts running program instructions at address 0 when it is powered up, and the program needs to be stored in Flash memory when the embedded system is operated.

You will also see the PIOA controller in the memory map. This is the PIO controller that we have been using to control our LEDs and read our switches. It has a number of registers starting at the address 0xFFFF F400 as shown in the memory map. We will now examine the PIO controller in more detail.

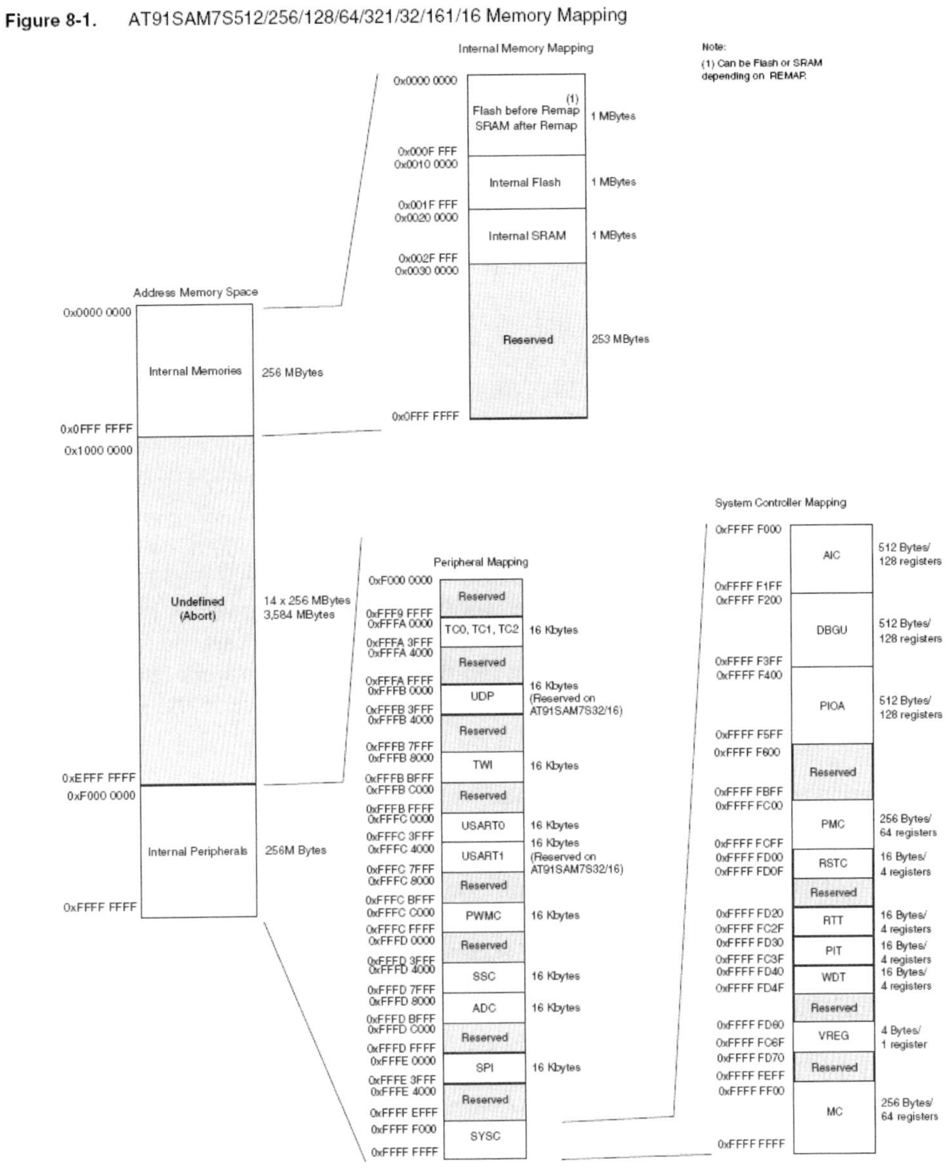

Figure 9.4: The AT91SAM7S Memory Map

9.5 A Closer Look at the PIO Controller

Chapter 15 of the AT91SAM7S microcontroller datasheet covers the Parallel Input/Output Controller (PIO). There are a lot of features that can be set up on the PIO controller such as glitch filtering, configurable pull-up resistors and more. We will only look at the basics of using the PIO controller to configure pins for use as input or output as we have already done in our first embedded programs.

Towards the end of chapter 27 of the datasheet in table 27-2, we can see the large number of registers available for configuring and using the PIO controller. The address of the start of the PIO controller's registers was shown to be 0xFFFF F400 from the memory map. In table 27-2 there is a column labelled "Offset" that shows the address offset from the PIO controller's base address of 0xFFFF F400. "Base address" refers to the first address in a range of addresses of a peripheral device.

Using the base address and offset address of the PIO controller, we can calculate the address of each register of the PIO controller by adding the two addresses together. Starting with the first register, the PIO_PER at offset 0x0000, we add this to the base address of 0xFFFF F400 and get 0xFFFF F400 for this register's address. The PIO_PER is the first PIO register and therefore falls on the PIO base address.

The next register, PIO_PDR, occurs on offset 0x0004. This is the very next 32-bit word address after the PIO_PER. The reason that it is not at offset 0x0001, being the next expected offset after offset 0x0000 is because we are dealing with 4 byte (32-bit) word addressing. Each address applies to 1 byte however, so to move between 4 byte registers, we have to increase the address by 4 bytes at a time. You will see that each of the registers are 4 bytes apart – one word apart on the 32-bit ARM. To work out the address of the PIO_PDR register, add the base address and offset address: 0xFFFF F400 + 0x0004 = 0xFFFF F404.

I have provided a header file, **`at91sam7s.h`**, as part of the template files that contains the addresses for hardware devices on the AT91SAM7S microcontroller family. If you open this file and scroll down or search for PIO, you will see each of the addresses for the PIOA registers. They correspond to the addresses in table 27-2 of the datasheet. Atmel have their own header files for these microcontrollers, but they are a lot harder to understand for the beginner C programmer. You will be able to see these header files used in the tutorial by James P. Lynch mentioned previously. The header file that I have provided contains each peripheral name and its corresponding de-referenced pointer address, making it easier to tie up the register names in the datasheet with the register names in the header file. This will be a lot easier for the beginner C programmer to understand.

If you scroll down from table 27-2 in the datasheet, you will come across details of each register from the table. We will examine each of the registers that we have used in our programs, namely PIO_PER, PIO_OER, PIO_ODR, PIO_PDSR, PIO_SODR and PIO_CODR. An additional register, PMC_PCER, from a different peripheral device was used when we read the switches on the PIO controller. We will examine why this was necessary shortly.

If you look back to chapter 6 on number systems in this book and at the example shown in **figure 6.2**, you will remember that we wrote a number to the port that had LEDs interfaced to it to switch the LEDs on and off. If you first wanted to switch only D1 on, you would write `0001B` to the port. If you then wanted to switch D4 on as well, you would then have to write `1001B` to the port and not `1000B`. If you wrote `1000B`, you would switch D4 on, but you would also switch D1 off. In your C program, you would then have to save the last value that you wrote to the port and individually set or clear the bit corresponding to the LED that you want to control. After modifying a single bit in a local copy (i.e. saved in a variable in your C program) of the port value, you would then write the new value to the port. This would leave the other bits of the port that were set unchanged, so if the LEDs on those bits were on or off, they would remain in that state and only the LED that you want to control will be changed. Controlling individual bits in a variable is possible by using bitwise operators in C. You have already seen the **AND** bitwise operator, there are others that we will look at in a later chapter.

To prevent the above scenario, the AT91SAM7S microcontroller has two registers to control the state of the port pins instead of one. One register is used for <u>setting</u> individual port pins by writing a 1 to each corresponding bit in this register. The other register is used for <u>clearing</u> individual port pins also by writing a 1 to each corresponding bit in this register. Setting, of course, refers to changing the state of the port pin to logic 1 and clearing refers to changing the state of the port pin to logic 0.

The PIO_SODR is used for setting pins on the microcontroller. From the detailed register information in the datasheet, we can see that writing a 1 to PIO_SODR sets the corresponding port pin and writing a 0 has no effect. With LEDs interfaced to PA0 and PA1 of the microcontroller's port, and only looking at the last four bits of the port, we would write `0001B` to switch PA0 to logic 1 and leave PA1, PA2 and PA3 in whatever state they were already in (logic 0 or logic 1). To switch both LEDs on at the same time we can write `0011B`. This will switch pins PA0 and PA1 to logic 1 and leave the other pins in their current state. Of course in this example we are assuming that the LEDs are wired to switch on with a logic 1 level, which is opposite to how the examples on the AT91SAM7S-EK board work. On this board we would be switching them off.

The PIO_CODR is used for clearing the pins of the microcontroller. In the datasheet we see that writing a 1 to this register clears the corresponding port pin and writing a 0 has no effect. To switch off the LED on PA0, write `0001B` to this register. This will set PA0 to logic 0 and leave the rest of the pins in their current state.

Before we can use the microcontroller pins as output (they actually default to inputs when the microcontroller powers up), we need to configure the pins that contain LEDs as output pins. We also need to set these pins up as I/O pins. The pins can be configured to connect one of the internal peripheral devices of the microcontroller to them or to connect the PIO controller to them (i.e. set them up as I/O pins). This is done using PIO_PER and PIO_OER.

When looking at the PIO_PER details in the datasheet, the following is stated:

0 = No effect.

1 = Enables the PIO to control the corresponding pin (disables peripheral control of the pin).

So to enable pins PA0 and PA1 as PIO pins, we write `0011B` to the PIO_PER. This sets up these two pins as PIO pins and leaves all other pins unchanged. In your C program this is the equivalent of writing 0x3 to the PIO_PER or if you write out all 32 bits 0x0000 0003. You can write it either way in your C program – the upper bits when writing 0x3 will always be zeros. You could even write the decimal number 3 in this case as hexadecimal 3 and decimal 3 are exactly the same, but writing in hexadecimal makes it clearer that you intend setting the two Least Significant Bits (LSBs) of the register.

The PIO_OER is used to set the PIO pins as outputs – they are set as inputs by default when the microcontroller is powered up. The pins are not set up as outputs by default for safety reasons. If they were set up as outputs and there was some device applying a logic level opposite to the one on the pin, a short-circuit would result.

From the datasheet it can be seen that setting bits in the PIO_OER enables the corresponding port pin to be used as an output.

If we want to use a pin as an input for reading switches, as an example, we need to set the pin up as an input. We do this by disabling the pin as an output. The pin should be set up as an input by default, but we set it up as an input anyway as a precaution. This is done by setting the corresponding bit in the PIO_ODR (PIO Output Disable Register).

To read back the logic state of the pin, we read back the state of the entire 32-bit port by reading the PIO_PDSR (PIO Pin Data Status Register). Because we have read the entire port, we then use the bitwise **AND** operator to isolate the state of our selected pin.

The datasheet states:
Reset value of PIO_PDSR depends on the level of the I/O lines. Reading the I/O line levels requires the clock of the PIO Controller to be enabled, otherwise PIO_PDSR reads the levels present on the I/O line at the time the clock was disabled.

The clock that the datasheet is referring to is controlled by another set of microcontroller memory mapped registers. In our program that read the switches and wrote to the LEDS was the following line:

```
PMC_PCER = 0x00000004;
```

This line is setting a bit in a register in order to enable the clock to the PIO controller. As the datasheet explains, we would not be able to read the status of the pins if this clock were not enabled. If you would like to read more about the clocks on the AT91SAM7S microcontrollers, refer to chapter 25 of the datasheet "Power Management Controller (PMC)".

9.6 Exercises

1. Write a program that flashes an LED on and off when a button on the embedded system is pressed. When the button is not pressed, the LED must remain off.

9.6.1 Solutions

1. `sw_blink`:

```
#include "at91sam7s.h"

void Delay(void);

int main(void)
{
    PMC_PCER = 0x00000004;      /* enable clock to PIO     */
    PIO_PER = 0x00008001;       /* enable uC pin as PIO    */
    PIO_OER = 0x00000001;       /* enable uC pin as output */
    PIO_ODR = 0x00008000;       /* pin used as input       */

    while (1) {
        if (PIO_PDSR & 0x00008000) { /* check if switch is pressed */
            PIO_SODR = 0x00000001;   /* switch out, switch LED off */
        }
        else {
            PIO_CODR = 0x00000001;  /* switch in, flash LED */
            Delay();
            PIO_SODR = 0x00000001;
            Delay();
        }
    }
}

void Delay(void)
{
    int del = 2000000;

    while (del--);
}
```

9.7 Summary

- Running programs from SRAM when developing and testing software is more convenient than running programs from Flash memory.
- An embedded system will need to run its software from non-volatile memory such as Flash memory after development when it is to be used for its intended purpose.
- The increment (++) and decrement (--) operators are used to add or subtract 1 from a variable.

- The bitwise **AND** operator (&) can be used to check if a single bit is set in a register or variable.

- The PIO controller on the AT91SAM7S microcontrollers is used to set the individual pins of the microcontroller up as inputs and outputs. It is also used to change the logic level on a pin set up as an output pin and read the state of a pin that is set up as an input pin.

- The addresses of memory and peripheral devices on a microcontroller can be found in the microcontroller's memory map.

10 The DBGU Serial Port

What you will do in this chapter:

- Find out how to connect your embedded system to the PC via the serial port
- Write a program that sends data out of the serial port
- Receive data on the PC from the embedded system
- Use the `sprintf()` function which is similar to the `printf()` function
- Send data from the PC to the embedded system via the serial port

This chapter is dedicated to getting communications between your embedded system and the PC working via the DBGU serial port on the AT91SAM7S microcontroller and the serial port on your PC. You will be able to send text messages from your embedded system to a terminal emulator program running on the PC and receive keyboard characters from the PC. Once the serial port is working, we can go on to learn more about the C language in the chapters that follow.

10.1 Hardware Requirements for PC to µC Serial Communications

Before we can program the embedded system's serial port, we need to set up hardware to connect the embedded system to the PC. The requirements are:

1 × serial port cable
1 × USB to RS-232 serial adapter if your PC does not have a serial port

10.1.1 Serial Cable

The serial cable only needs to link up three of the RS-232 lines on the embedded system to the PC. When linking up the RS-232 serial ports, the transmit (Tx) line of the embedded system's port must be connected to the receive (Rx) line of the PCs port. The receive line (Rx) of the embedded system must be connected to the transmit (Tx) line of the PC. The ground pin (GND or 0V) of each system must be connected.

The PC uses a 9 pin male D-type connector for its serial port. The pins of this port that we are going to use are shown on the left in **figure 10.1**.

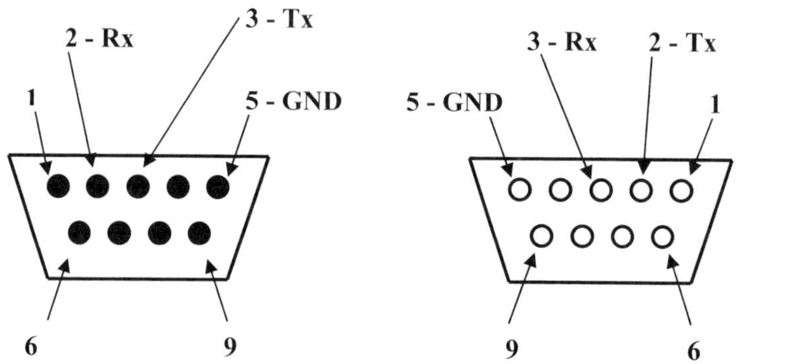

9 pin D-type male (plug) **9 pin D-type female (socket)**

Figure 10.1: D-type RS-232 Connectors

It is possible that your embedded system has an RS-232 connector the same as the PC or a 9 pin female connector also shown on the right in **figure 10.1**. In either case, always check how your embedded system serial port is wired to the connector by looking at the schematic diagram or documentation. The pin numbers shown in the figure are standard configurations so your embedded system should match one of them.

You will need to make or buy a cable with wiring that depends on which connector is used on your embedded system. If the embedded system contains a male plug as shown, you will need a cable that crosses over the Tx and Rx lines. If your embedded system has the female plug as shown, you will need a straight through cable that connects the pins one to one and will therefore also join the Tx and Rx lines. **Figure 10.2** shows the two possibilities.

Both cables are effectively joining the Tx line of the PC to the embedded systems Rx and the Rx line of the PC to the embedded systems Tx. The connectors on the cable will obviously have to be opposite gender to the connector that they are plugging into.

The AT91SAM7S-EK board that I am using will use the cross-over cable configuration as the RS-232 ports on this board are configured the same as the PC. Another name for this type of cable is "null-modem cable". If you are going to buy a pre-made cable, this is what to ask for. The null-modem cable will have the other RS-232 pins wired as well, but this is not a problem.

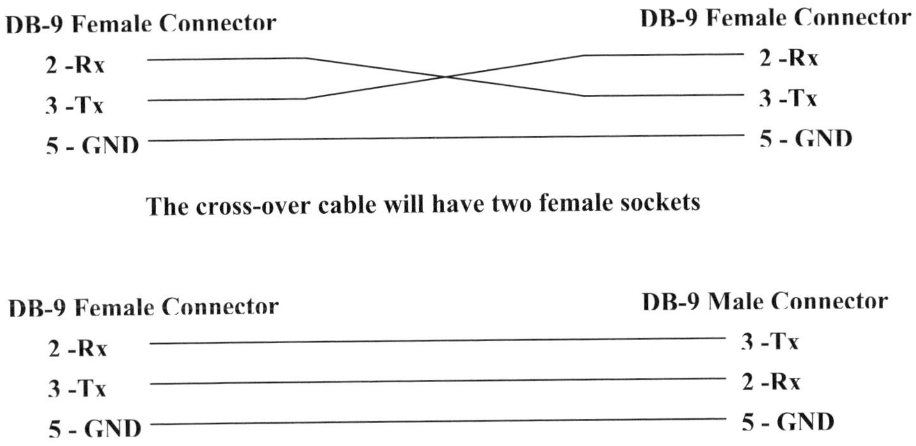

The cross-over cable will have two female sockets

The straight through cable will have one female socket and one male plug

Figure 10.2: Two Possible RS-232 Cable Configurations

10.1.2 USB to RS-232 Serial Adapter

If your PC does not have a RS-232 port already on it like most of the notebook or laptop computers available today, you will need to buy a USB to RS-232 converter. Any computer shop should be able to supply you with one of these devices. It may be worthwhile to buy one that has two RS-232 serial ports on it. The device is simply plugged into the PC's USB port and will come with drivers for configuring it if necessary.

Plug the USB to RS-232 port converter into the PC and configure it now according to the instructions that were supplied with it. We will be using it with our next program. If your PC already has an RS-232 port, you don't need to configure anything.

10.2 Programming the Serial Port

With the hardware ready, we can now create a program to initialise the serial port and transmit some data from it to a terminal emulator program running on the PC. We will be using a particular serial port on the microcontroller called the DBGU port. Your embedded system needs to have this port interfaced to an RS-232 level converter chip and to an external DB-9 connector.

Create a new embedded project directory called **`serial_tx`** and copy your template files to it. Open **`main.c`** for editing and add the following lines of code:

```
#include "at91sam7s.h"

#define DBGU_115200     26
```

```
void DBGUInit(unsigned int baud);
void DBGUTxMsg(char* msg);

int main(void)
{
    DBGUInit(DBGU_115200);

    DBGUTxMsg("Hello from DBGU port!\r\n");
    while (1) {
    }
}

void DBGUInit(unsigned int baud)
{
    /* connect the DBGU UART to PA9 & 10 */
    PIO_ASR = 0x00000600;       /* select the DBGU */
    PIO_PDR = 0x00000600;       /* DBGU controls the pins */
    /* setup the DBGU UART */
    DBGU_BRGR = baud;           /* set baud rate */
    DBGU_MR = 0x00000800;       /* no parity, normal mode */
    DBGU_CR = 0x00000050;       /* enable tx and rx */
}

void DBGUTxMsg(char* msg)
{
    int ch_num = 0;

    /* send characters until null received */
    while (msg[ch_num] != 0) {
        while (!(DBGU_SR & 0x00000002));
        DBGU_THR = (unsigned int)msg[ch_num];
        ch_num++;
    }
}
```

The **`serial_tx`** program calls a function to initialise the DBGU serial port of the AT91SAM7S microcontroller. The initialisation consists of connecting the DBGU serial port device that is in the microcontroller chip to the port pins PA9 and PA10 of the microcontroller, setting up the baud rate and other parameters of the DBGU port and then enabling the DBGU port to transmit and receive. The baud rate setting relies on the system having an 18.432MHz crystal and the system clock being set up as in the assembly language start up template file. We will look at the hardware in greater detail after running the program.

A function is then called to transmit a text message out of the serial port. Program flow drops into a **`while(1)`** loop that does nothing after this.

Build this project and then plug in your hardware as follows:

1. If you are using a USB to serial port adapter, make sure that it is plugged into the PC and ready for use.

2. Plug the RS-232 cable into the serial port of the PC or USB to serial converter and then into the DBGU serial port of your embedded system.

3. Start the Windows HyperTerminal program: **start → All Programs → Accessories → Communications → HyperTerminal (figure 10.3)**

4. In the **Connection Description** dialog box that pops up, type a name to describe the connection (**figure 10.4**) and then click the **OK** button.

5. In the **Connect To** dialog box, select the COM port that you have connected your embedded system into. Use the **Connect using:** drop down box to select it (**figure 10.5**). Click the **OK** button.

6. In the **COM Properties** dialog box (**figure 10.6**) choose **115200** in the **Bits per second:** drop down box and **None** in the **Flow control:** drop down box. The rest of the items must remain at their default values as shown in **figure 10.6**. Click the **OK** button.

7. Connect your embedded system to the PC via the USB cable for programming and power up the board.

8. Using the SAM-BA PC program, load the binary file from the `serial_tx` project into the microcontroller's SRAM memory as you have previously done.

9. Run the program.

If all went well, you should get a message displayed in your HyperTerminal program's window as shown in **figure 10.7**.

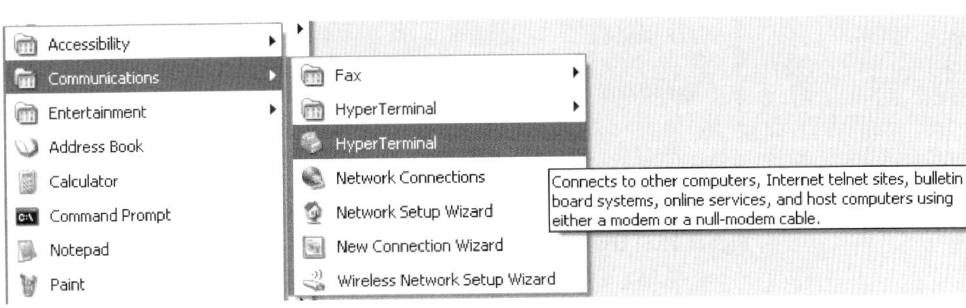

Figure 10.3: Starting the HyperTerminal Program

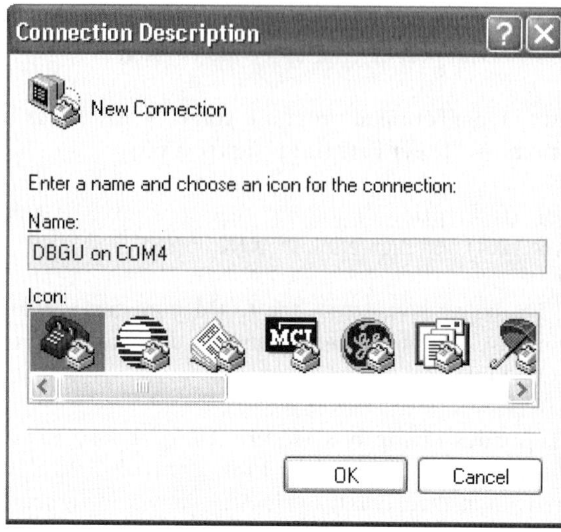

Figure 10.4: HyperTerminal's Connection Description Dialog Box

Figure 10.5: HyperTerminal's Connect To Dialog Box

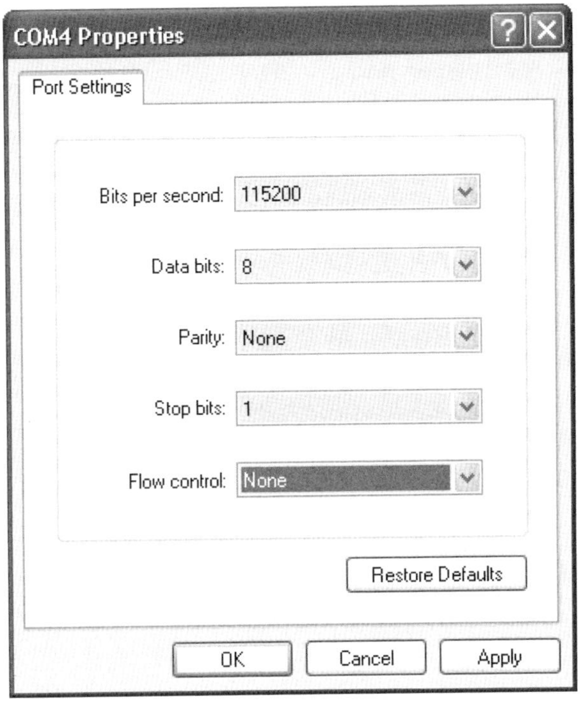

Figure 10.6: HyperTerminal's COM Properties Dialog Box

If your program failed, some of the possibilities that made it fail are:

1. Incorrectly wired serial cable.
2. Connecting to the wrong COM port with HyperTerminal.
3. Not setting up the baud rate or other communication parameters in HyperTerminal correctly.
4. Incorrect hardware on embedded system: Not connected to the DBGU port of the AT91SAM7S microcontroller, incorrect crystal or oscillator setup on the board. The oscillator would be set up correctly if you are using the template files and your board has an 18.432MHz crystal. Note that on the AT91SAM7S-EK there are two serial ports; make sure that you choose the DBGU serial port if you are using this board.

When you close HyperTerminal, you will be asked firstly if you want to disconnect from the port, select **Yes** in the dialog box. Next you will be asked if you want to save the connection. It is a good idea to save it as you won't have to set up the communication parameters when using HyperTerminal again, so click **Yes** in the dialog box. The connection will be available under **Communications → HyperTerminal → <the name that you selected in figure 10.4>** as shown in **figure 10.3** (the pop out HyperTerminal menu above the HyperTerminal program that you clicked to start HyperTerminal).

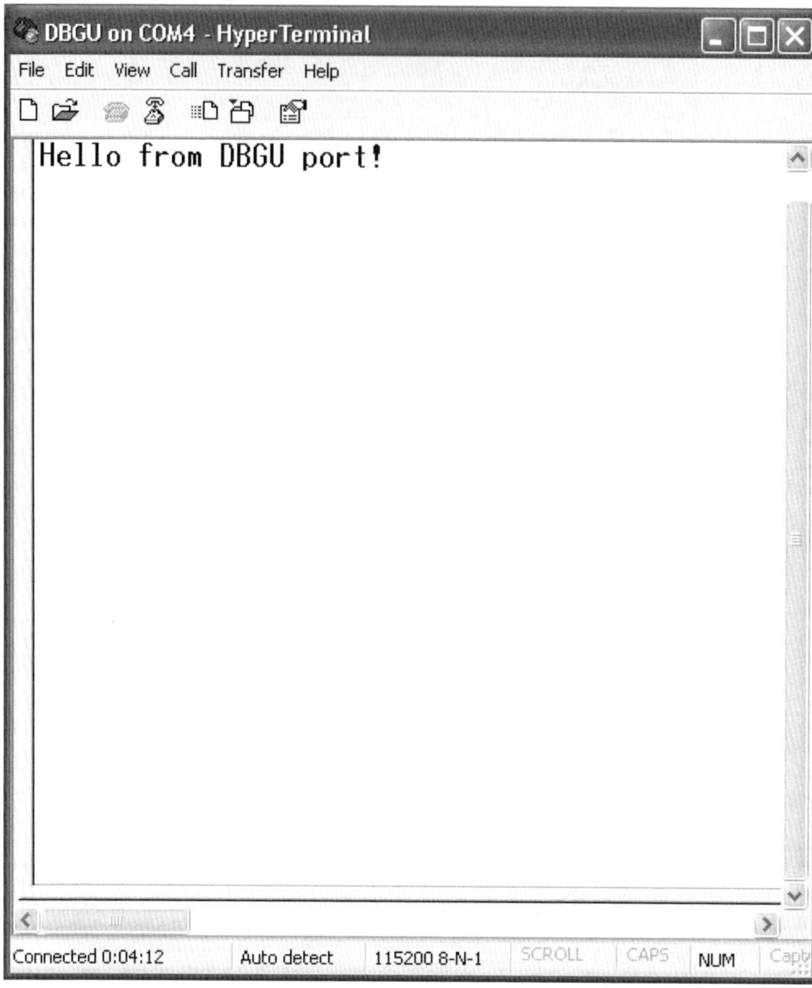

Figure 10.7: The Result of Running the `serial_tx` Program

Figure 10.8 shows the hardware setup for using the DBGU serial port with HyperTerminal running on the PC.

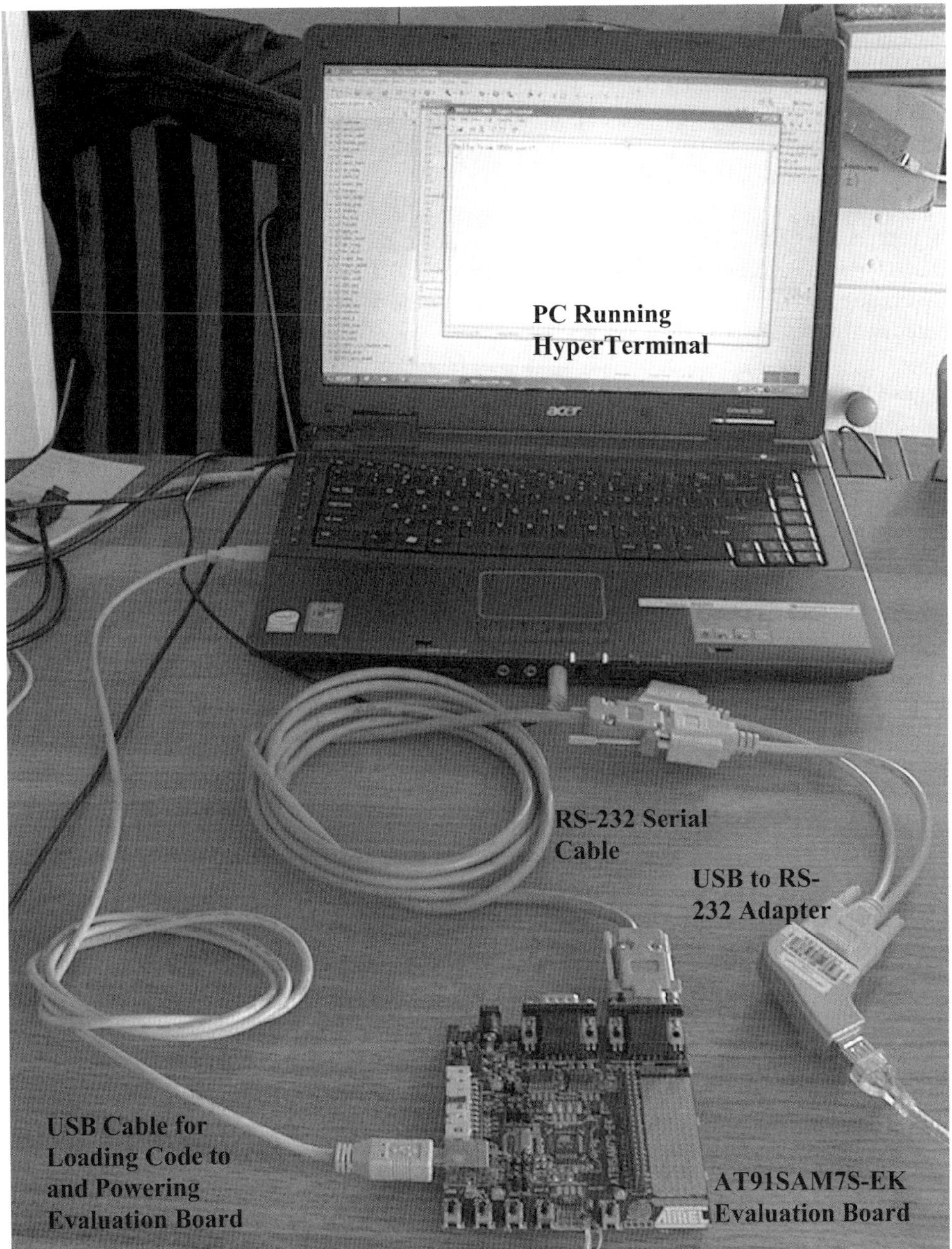

Figure 10.8: Hardware Setup used with the DBGU Serial Port

10.2.1 About the DBGU Serial Port Hardware

After programming the PIO port you should now understand that programming an embedded system in C relies on knowing the C language as well as knowing how the piece of hardware works that you are programming. If you don't know how the hardware works, you will not know what to write to the hardware registers to configure and use it. To learn how to use the hardware we must refer to the microcontroller datasheet, but any source code that can be found that is already written to use the hardware will also be a big help and time saver (e.g. from source code samples and projects on the Internet, application notes and programs from Atmel).

If you look at the AT91SAM7S microcontroller's memory map, you will see that the DBGU port is memory mapped at address 0xFFFF F200. Chapter 26 of the datasheet covers the Debug Unit (DBGU) which contains the DBGU serial port as well as some other hardware devices.

To use the DBGU port as an RS-232 serial port, an external RS-232 level converter chip must be interfaced to the transmit and receive pins of the DBGU. This will already be in place on your embedded system if the hardware designer decided to wire up the DBGU port as an RS-232 port.

The pins used for the DBGU serial port are port pins PA9 and PA10 of the microcontroller. The AT91SAM7S is designed in such a way that these pins can be used for one of three possible functions. They could be set up as input or output pins by writing to internal registers that connect them to the PIO controller. There are then two possible hardware devices that they could be connected to, one of them being the DBGU serial port. Port pin PA9 can be internally connected to function as DRXD which is the DBGU receive pin of the DBGU serial port and port pin PA10 can be internally connected to function as DTXD which is the transmit pin of the DBGU serial port.

In order to use the serial port, it must be internally connected to the port pins PA9 and PA10. It is necessary first to write to the PIO_ASR (A Select Register) which selects peripheral A – the DBGU port. Table 10-3 in chapter 10 of the datasheet shows which peripherals are peripheral A and which are B. The table also shows that the DRXD pin of the DBGU serial port can be connected to pin PA9 and DTXD to pin PA10. The PIO_ASR is one of the PIO controller's registers. The PIO controller is used to internally wire up the pins as I/O pins or pins from one of the internal peripherals.

After the DBGU serial port is selected for use on pins PA9 and PA10, the PIO_PDR (PIO Disable Register) is written to in order to disable the PIO controller from controlling these two pins as general purpose inputs and outputs. The DBGU serial port is now connected and ready for use.

Three DBGU registers are required in order to set up the DBGU port. The DBGU_BRGR is used to set up the baud rate of the serial port. The baud rate sets the speed that serial port communicates at. In our example program, the embedded system serial port was set to 115200 baud and HyperTerminal was also set to 115200 baud so that they could communicate. Chapter 26 of the AT91SAM7S datasheet explains how to calculate the baud rate.

The DBGU_MR (Mode Register) is used to set the parity, another serial communication setting, to none. Finally, DBGU_CR (Control Register) is used to enable the transmitter and receiver of the DBGU serial port. We will see how each of these registers is written to when we analyse the **serial_tx** program next.

10.2.2 How the serial_tx Program Works

The basic operation of the program has already been described, we will now examine the initialisation and transmit functions that are called in **main()**.

Figure 10.9 shows the code that we find inside the **DBGUInit()** function.

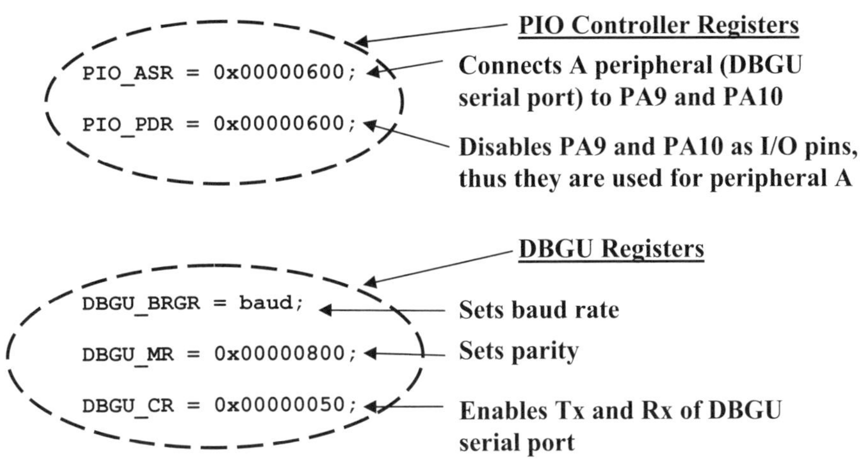

Figure 10.9: The DBGUInit() Function

Each of the hardware registers that are set up were described in the previous section. To understand which bits in each register are being set, you will need to refer to the detailed description of each register in the datasheet. The two PIO registers are described at the end of chapter 27 in the datasheet. The three DBGU registers are described at the end of chapter 26 in the datasheet. The baud rate was defined using the **#define** directive. Normally when a function that initialises a serial port is written, a number of different baud rates are defined in this manner and placed in a header file. The same function can then be used to easily set up a number of standard baud rates easily, depending on what the application requires.

The **DBGUTxMsg()** function is used to transmit a text string out of the serial port. This function is passed a pointer to the text string and this is how the function knows what to transmit. The function transmits one character of the string at a time until it reads a value of 0. This tells the function that the end of the string has been reached. Unfortunately to fully understand the C syntax here, we will need to look further at pointers and also cover another topic – arrays. We will be looking at these topics in detail in a separate chapter, but for now, let's see how we use the serial port's registers to transmit each byte.

The two statements in the **DBGUTxMsg()** function used for transmitting a byte of data are :

```
while (!(DBGU_SR & 0x00000002));

DBGU_THR = (unsigned int)msg[ch_num];
```

The first statement checks the TXRDY bit in the DBGU status register, DBGU_SR, by using the bitwise **AND** operator. You will see that this bit is bit number 1 in the DBGU_SR (the second bit from the right of the register) by looking in the datasheet (**figure 10.10**).

31	30	29	28	27	26	25	24
COMMRX	COMMTX	–	–	–	–	–	–
23	**22**	**21**	**20**	**19**	**18**	**17**	**16**
–	–	–	–	–	–	–	–
15	**14**	**13**	**12**	**11**	**10**	**9**	**8**
–	–	–	RXBUFF	TXBUFE	–	TXEMPTY	–
7	**6**	**5**	**4**	**3**	**2**	**1**	**0**
PARE	FRAME	OVRE	ENDTX	ENDRX	–	TXRDY	RXRDY

Figure 10.10: The DBGU Status Register (DBGU_SR)

If this bitwise **AND** operation results in true i.e. TXRDY is set, then this tells us that the transmitter of the DBGU serial port is ready for us to transmit a byte. Because it is true, we would then be stuck in the **while** loop. The **not** operator is used to invert the result of the bitwise operator evaluation. We will therefore break out of the loop when the transmitter is ready. If the transmitter is not ready, a 0 is read from the TXRDY bit and when modified with the **not** operator the result will evaluate to true and we will stick in the loop and keep checking the TXRDY bit until it is set. In other words what we are doing here is waiting in the loop until the transmitter is ready.

The **while** loop written in this way:

```
while (!(DBGU_SR & 0x00000002));
```

is the same as writing it this way (an empty loop body):

```
while (!(DBGU_SR & 0x00000002))
{
}
```

which is the same as writing it this way:

```
while (!(DBGU_SR & 0x00000002)) {}
```

Because there is nothing to do in the loop body, the loop expression is continually being evaluated as fast as this line of code can be run by the microcontroller.

The second line of the function shown above, writes the character that is to be transmitted to the Transmit Holding Register (DBGU_THR). When the character has been written to this register, it will be transmitted by the DBGU port serially to HyperTerminal running on the PC. The line is shown here:

```
DBGU_THR = (unsigned int)msg[ch_num];
```

If you look at the DBGU_THR in the datasheet, you will see that the register is 32 bits wide. Only the least significant byte of this register is used, however, because the serial port only transmits a byte at a time. Whatever gets written to the top 24 bytes is ignored. The notation of the above statement requires some explanation. A single character in the string is being accessed in this line by using array notation. This will be covered in a later chapter as already mentioned. Another C element is used here called a “cast”. This is the **`(unsigned int)`** that you see in brackets. The character from the string is cast from a **`char`** to an **`unsigned int`**. Because we are writing an 8 bit character to a 32 bit register, the compiler will complain because there is a size difference. Casting converts the 8 bit character to an unsigned integer, in this case, by copying it to the lower byte of an internal variable before sending it to the DBGU_THR register. We will look at casts again later, so don’t worry if you don’t fully understand it at this stage.

10.3 Using the `sprintf()` Function

The **`sprintf()`** function is almost equivalent to the **`printf()`** function that we used in our PC C programs. The **`printf()`** function prints to the standard I/O device on the PC. The **`sprintf()`** function (string print formatted) prints to a string, also known as an array of characters.

There is no string type in C, we must create an empty array of characters in which **`sprintf()`** can write a string. The following line will create an array of 40 characters in a C program:

```
char my_str[40];
```

This creates space in memory for 40 consecutive characters that we can access in our program. As mentioned, we will cover strings and arrays in a separate chapter. With the above line, we are ready to use the **`sprintf()`** function.

The **`string_tx`** program will demonstrate the use of the **`sprintf()`** function. This program can be created by copying the previous program to a new directory and then modifying it as we are going to reuse the serial port functions.

```
#include <stdio.h>
#include "at91sam7s.h"

#define DBGU_115200     26
void DBGUInit(unsigned int baud);
void DBGUTxMsg(char* msg);

int main(void)
{
    char my_str[40];

    DBGUInit(DBGU_115200);
    sprintf(my_str,
        "There are %lu bytes in the array\r\n", sizeof(my_str));
    DBGUTxMsg(my_str);

    while (1) {
    }
}
```

The DBGU port functions are not included in this listing as they are exactly the same as the previous program. They must however be present in **main.c** in order for the program to compile.

You will notice something familiar right at the top of the program, namely the inclusion of the **stdio.h** header file. This header file is necessary as it contains the function prototype for the **sprintf()** function.

Inside **main()**, an array of characters is defined in order for **sprintf()** to store our string. We can write a string that is up to 39 characters long to this array, you will see why only 39 and not 40 in the chapter on strings. The DBGU port is initialised as in our previous program. The **sprintf()** function is then called. The function call looks almost the same as the **printf()** function, except that the first parameter that we pass to it is the name of the **my_str** array. This function then prints the string between quotes to the array. In this example, the size of the array is substituted into the string using the C **sizeof** keyword which returns the size in bytes of the array.

Finally the formatted string that is now stored in the **my_str[]** array is sent out of the serial port to be displayed on the HyperTerminal terminal emulator. Be sure to start up HyperTerminal before you load and run this program so that you can capture the output from the embedded systems serial port.

When sending strings to HyperTerminal, we need to send two escape sequences, **\r** and **\n**, to move the cursor to the beginning of the next line in this program. If only the newline character was sent, the cursor would move to the next line, but would remain in its current horizontal position. The carriage return character moves it to the beginning of the line.

Something to note when using the **sprintf()** library function is the size of the binary output file that is produced when building the program. In the **serial_tx** program that does not use this function, the size of the binary file that the compiler generated on my system was 668 bytes. Compare this to the size of the binary output file of the **string_tx** program which is 38.4k bytes. You can see that using **sprintf()** in a program uses a huge amount of memory. Microcontrollers with smaller amounts of memory would obviously not be able to run this function. Although this function uses more than half of the available SRAM on the microcontroller, it would not be such a problem when this program is loaded into and run from Flash memory on the AT91SAM7S256 microcontroller with its 256k bytes of Flash memory. A more efficient function can be written if limited formatting capabilities are required. If, for example, your program only needed to print out decimal numbers in ASCII format, you could write a small function to do this.

10.4 Receiving Data on the Serial Port

We can send data from HyperTerminal to our embedded system. In order to see if a byte has been received by our embedded system, we need to continually check a bit in one of the DBGU registers that will be set if the DBGU serial port has received a byte. If, when we read this bit it is set, we can then read the byte that was sent out of another of the DBGU registers. The **serial_rx** program will demonstrate how to receive a byte when we type a keyboard character into HyperTerminal and it is sent to our embedded system.

```
#include "at91sam7s.h"

#define DBGU_115200      26
void DBGUInit(unsigned int baud);
char DBGUCheckForRxData(void);
char DBGURx(void);

int main(void)
{
    char rx_byte;

    PIO_PER = 0x00000001;         /* enable uC pin as PIO */
    PIO_OER = 0x00000001;         /* enable uC pin as output */
    DBGUInit(DBGU_115200);

    while (1) {
        if (DBGUCheckForRxData()) {         /* is a byte received? */
            rx_byte = DBGURx();             /* yes, get the byte */
            if (rx_byte == 'a') {           /* a received */
                PIO_CODR = 0x00000001;      /* switch LED on */
            }
            else if (rx_byte == 'b') {      /* b received */
                PIO_SODR = 0x00000001;      /* switch LED off */
            }
        }
    }
```

```
}

char DBGUCheckForRxData(void)
{
    /* check if character has been received */
    if (DBGU_SR & 0x00000001) {
        return 1;
    }
    return 0;
}

char DBGURx(void)
{
    return (char)DBGU_RHR;      /* read the character */
}
```

The **DBGUInit()** function has been left out of this listing as it is the same function that you have already used. It must be inserted into your **main.c** source file.

Before building this project, you will need to modify it to use a single LED on your board. This program controls an LED that is attached to PB0 on the AT91SAM7S. Change all the writes to the PIO registers (they occur in 4 places in **main()**) in this program to use the port pin that controls the LED on your embedded system.

Start up HyperTerminal as before and then load and run this program on your embedded system. In HyperTerminal, pressing the 'a' and 'b' keys on the keyboard will send the 'a' and 'b' characters to the embedded system via the serial port. The program will switch the LED on and off when the 'a' and 'b' characters are received. Make sure that your Caps Lock key is not active. If it is, you will send a capital 'A' and 'B' and the program will not work.

Inside **main()**, the microcontroller pin that controls the LED is set up first. The DBGU serial port is then initialised as we have done previously. Inside the **while(1)** loop, an **if** statement continually checks to see whether a character has been received on the DBGU serial port. It does this by calling the **DBGUCheckForRxData()** function. This function simply checks a single bit in one of the DBGU registers which will be set if a character has been received on the serial port. The function returns 1 if a character has been received, or else it will return 0.

When a character is received, the **DBGURx()** function is called which will return the character that was received on the serial port. The character is stored in the **rx_byte** variable. The **DBGURx()** function obtains the character that was sent to the serial port by reading the **DBGU_RHR** register.

After the serial port character has been received, a check is done to see if it was an 'a' or a 'b' that was sent. If the character was an 'a', the LED is switched on. If the character was a 'b', the LED is switched off. If any other character was received, the program does nothing with it. The program will then go back to continually checking for another character on the serial port.

From this program, it can be seen how the DBGU serial port works. The RXRDY bit in the DBGU_SR (DBGU Status Register) will always be read as 0, until a character has been received by the DBGU serial port. If a character has been received by the serial port, it will first be moved into the DBGU_RHR (DBGU Receive Holding Register) and then the RXRDY bit will be set to 1. You can take a look at these registers in chapter 26 of the datasheet. The RXRDY bit is the LSB (bit 0) in the DBGU_SR. The datasheet also states that: **The bit RXRDY is automatically cleared when the receive holding register DBGU_RHR is read.** Thus, after getting the byte out of the DBGU_RHR in our C program, the RXRDY bit will be cleared to 0 and we are ready to check for the arrival of a new byte on the DBGU serial port.

10.5 Exercises

1. Write a program that sends a message to HyperTerminal from your embedded system when a character is sent from HyperTerminal to your embedded system. The message should tell the user which character was sent.

10.5.1 Solutions

1. **`which_char`**:

```
#include <stdio.h>
#include "at91sam7s.h"

#define DBGU_115200      26
void DBGUInit(unsigned int baud);
void DBGUTxMsg(char* msg);
char DBGUCheckForRxData(void);
char DBGURx(void);

int main(void)
{
    char my_str[40];
    char rx_byte;

    DBGUInit(DBGU_115200);
    while (1) {
        if (DBGUCheckForRxData()) {      /* is a byte received? */
            rx_byte = DBGURx();          /* yes, get the byte */
            sprintf(my_str, "You pressed %c\r\n", rx_byte);
            DBGUTxMsg(my_str);
        }
    }
}
```

This program uses four functions that we have already used in this chapter, so copy them from the example programs. The functions are not included in the above listing, but the function prototypes are, so check the function prototype names to see which functions you need.

10.6 Summary

- The serial port on the embedded system allows us to easily send and receive text to and from a PC. Note that the serial port is not limited to sending ASCII text, but can send any byte value.

- A terminal emulator such as HyperTerminal running on a PC can be used to communicate with an embedded system.

- The **`sprintf()`** function is very similar to the **`printf()`** function and allows us to insert numbers and other data into a string for transmission over the serial port.

11 Previous C Topics Revisited

What you will do in this chapter:

- Use a new set of template files that include a serial port driver
- Look at the previous C topics covered in chapters 1 to 7 in an embedded context
- Expand on the previous C topics

11.1 Serial Port Driver

A new set of template files called **`RAM_template2`** have been provided for you and can be found in the zipped source code file. Refer to appendix B for details.

The template files now include an extra C source file and header file called **`dbgu.c`** and **`dbgu.h`**. These files contain the driver for the serial port. A driver is a set of functions that can be used to control a piece of hardware. In this case, the driver is a set of C functions that can be called in a program to initialise and operate the DBGU serial port. You will recognise the initialisation routine and other functions as ones that we have already used in the previous chapter.

The template files have been arranged into subdirectories to make them easier to use. The **`startup`** subdirectory contains the assembly language start-up file as well as the linker directive file. The **`Drivers`** subdirectory contains the **`dbgu.c`** and **`dbgu.h`** driver files as well as **`at91sam7s.h`** and the **`Code`** subdirectory contains **`main.c`**. The root directory of the template files contains the make file. The make file has been set up to look for the various source files in the subdirectories. I have chosen this structure in order to divide the project into logical blocks. All the files that are generated by the programming tools will be created in the root directory of the template. This makes it easy to create a new project that is based on a project that you have already made. All you need to do is delete the files that the software tools generated, copy the entire project to a new directory and rename the project in the make file. It is also easier to find the files that make each project unique, as well as the files that are used in every project, such as the drivers. **Figure 11.1** shows the template structure.

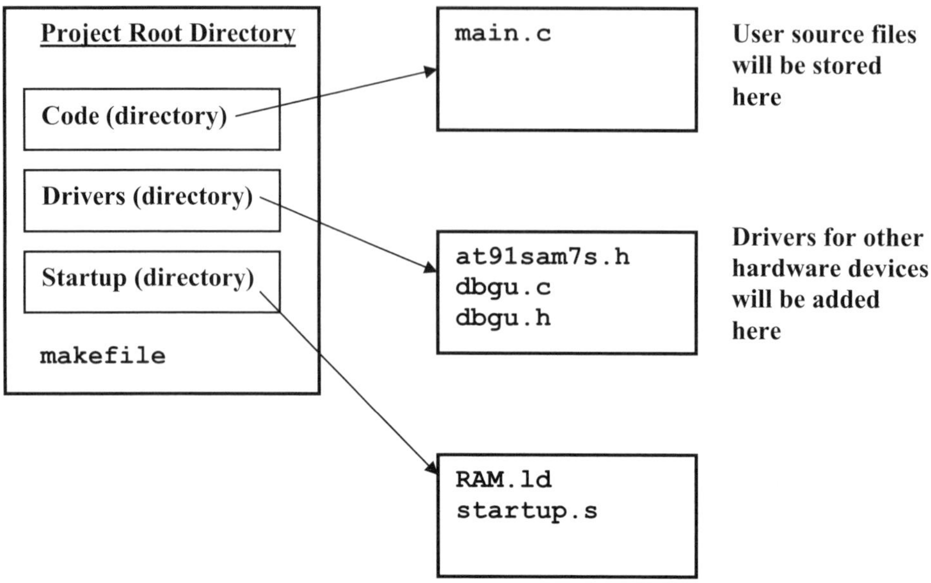

Figure 11.1: The New Template Files and Directory Structure

Now that we have our embedded system, serial port driver and a serial port link to connect the embedded system to the PC, we are ready to expand our knowledge of C programming that we have already covered and then look at some new C programming topics in the chapters that follow.

11.2 Format Specifiers

Table 11.1 shows the format specifiers in C. You have already used some of them. The **formats** program makes use of some of the new format specifiers.

```
#include <stdio.h>
#include "dbgu.h"

int main(void)
{
    char str[50];

    DBGUInit(DBGU_115200);
    /* characters and strings */
    sprintf(str, "%s %c\r\n", "This is the letter", 't');
    DBGUTxMsg(str);
    /* integers */
    sprintf(str, "Ints:%d, %i, %u\r\n", -1000, -123, 958);
    DBGUTxMsg(str);
    sprintf(str, "Oct:%o, hex:%x, HEX:%X\r\n", 191, 191, 191);
    DBGUTxMsg(str);
    /* floating point numbers */
    sprintf(str, "%f, %e, %E\r\n", 45.345, 45.345, 45.345);
```

```
    DBGUTxMsg(str);
    while (1);
}
```

Format specifiers work the same with **sprintf()** and **printf()**. **%s** and **%c** are used to insert a string and character into a string. You have already seen **%d**, **%i** does the same thing and **%u** works with unsigned integers. **%x** and **%X** are for printing hexadecimal numbers. **%o** prints out a number in octal format. Octal is another number system like binary or hexadecimal, but uses base 8. **%f** is familiar to you and is used to print a floating point number. **%e** and **%E** print a number in exponential notation, which is a notation used by scientists and engineers.

Notice again that the **while(1)** loop will stop the program flow from exiting **main()**. This is necessary as there is no operating system to return to and it prevents the microcontroller from trying to execute instructions from memory outside of **main()**.

Format Specifier	**Meaning**
%c	character
%s	string
%d, %i	decimal integer
%u	decimal unsigned integer
%o	octal unsigned integer
%x	hexadecimal unsigned integer, lowercase a to f
%X	hexadecimal unsigned integer, uppercase A to F
%f	floating point number
%e, %E	exponential notation

Table 11.1: C Format Specifiers

11.3 Escape Sequences

You have already seen the newline, tab and carriage return escape sequences. **Table 11.2** shows the other possible escape sequences in C.

Escape Sequence	**Meaning**
\a	alert (bell character)
\b	backspace
\f	formfeed
\n	newline
\r	carriage return
\t	horizontal tab
\v	vertical tab
\\	backslash

`\?`	question mark
`\'`	single quote
`\"`	double quote
`\ooo`	octal number
`\xhh`	hexadecimal number

Table 11.2: C Escape Sequences

An escape sequence always begins with a backslash. This gives the backslash character special meaning in C. To print the backslash character, we therefore need to use an escape sequence. This is the double backslash (\\) that you see in the table.

The same applies to the single and double quotes. These characters have special meaning in C, so to print one of these characters out, we need an escape sequence. For example the following statement:

```
printf("I said, "Hello"");
```

would cause a compile error because the compiler sees **`"I said, "`** as a string. **`Hello`** is not part of the string, but falls outside of it. We need to use escape sequences to be able to print this text:

```
printf("I said, \"Hello\"");
```

Characters can also be inserted directly into a string by using the hexadecimal number escape sequence. This will insert the equivalent ASCII character from the ASCII table in place of the hexadecimal number.

It is possible to use the hexadecimal number escape sequence to print characters that you would not be able to type into your C source code, such as the graphical characters from the ASCII table.

The **`escape`** program demonstrates the use of some of the new escape sequences. The program shows two ways of inserting quotes into a string, how to insert characters into a string from the ASCII table and how to insert graphical characters into a string from the ASCII table.

```
#include <stdio.h>
#include "dbgu.h"

int main(void)
{
    char str[50];

    DBGUInit(DBGU_115200);

    /* printing quotes in a string */
    sprintf(str, "I said, %cHello%c\r\n", '"', '"');
    DBGUTxMsg(str);
    sprintf(str, "I said, \"Hello\"\r\n");
    DBGUTxMsg(str);
    /* using hex number escape sequences */
```

```
    sprintf(str, "%c%c%c\r\n", 'A', 'B', 'C');
    DBGUTxMsg(str);
    sprintf(str, "\x41\x42\x43\r\n");
    DBGUTxMsg(str);
    /* using hex numbers to print graphical characters */
    sprintf(str, "\xC9\xCD\xCD\xCD\xCD\xCD\xBB\r\n");
    DBGUTxMsg(str);
    sprintf(str, "\xBA ARM \xBA\r\n");
    DBGUTxMsg(str);
    sprintf(str, "\xC8\xCD\xCD\xCD\xCD\xCD\xBC\r\n");
    DBGUTxMsg(str);

    while (1);
}
```

The program first prints out quotes using format specifiers and character constants. The same string is then printed out using escape sequences which directly inserts the quotes into the string, no need for format specifiers.

An example is then given of printing out the characters A, B and C first using character constants and then using the hexadecimal number escape sequence. The hexadecimal number is used to print out the same characters by using the correct number from the ASCII table.

Finally the program prints out some graphical characters from the ASCII table that you would not be able to type into your C source code. The output of the program is shown in **figure 11.2**. Refer to Appendix A for the ASCII table.

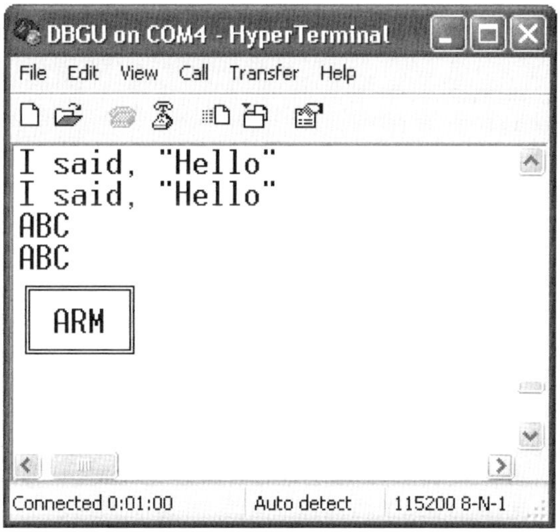

Figure 11.2: The Output of the escape Program

11.4 Loops

We have only seen one type of loop in C so far, the **while** loop. We will look at two more C loops, the **do while** loop and the **for** loop, but let's first look at the **break** and **continue** statements used in a **while** loop. These statements can be used in the other types of loops as well.

11.4.1 A while Loop that uses break and continue

The **brk_cont** program demonstrates how the **break** and **continue** statements work.

```
#include <stdio.h>
#include "dbgu.h"

int main(void)
{
    int count = 0;
    char str[50];

    DBGUInit(DBGU_115200);

    while (count < 10) {
        count++;
        if (count == 5)
            break;
        sprintf(str, "Count 1 = %d\r\n", count);
        DBGUTxMsg(str);
    }
    count = 0;
    while (count < 10) {
        count++;
        if (count == 5)
            continue;
        sprintf(str, "Count 2 = %d\r\n", count);
        DBGUTxMsg(str);
    }

    while (1);
}
```

In this program, the **break** statement is demonstrated first. In the first **while** loop, a count is incremented at each pass through the loop and the result is sent to the serial port. When the value of the count reaches 5, the break statement is run. This statement breaks out of the loop altogether and will no longer run any of the code in the loop or evaluate the loop expression again. Program flow continues below this **while** loop.

The **continue** statement is used in the second **while** loop. In this loop, the **continue** statement is run when the count reaches a value of 5. The **continue** statement causes program flow to continue at the top of the loop and skip all of the statements below it. In this program, the printing out of the count value to the serial port is skipped when the count value is equal to 5. This is because the two statements below the **continue** statement are not run at all, instead the loop expression is evaluated and then the count incremented at the top of the loop.

11.4.2 The do while Loop

The **do while** loop is very similar to the **while** loop, except that it is always run once, no matter what the loop expression evaluates to. This is because the loop expression is evaluated *after* each pass through the loop. You can use this kind of loop every time that you want the code in the loop to be run at least once, no matter what the loop expression result is. The **do_while** program shows how this works.

```
#include <stdio.h>
#include "dbgu.h"

int main(void)
{
    int count = 0;
    char str[50];

    DBGUInit(DBGU_115200);
    do {
        count++;
        sprintf(str, "Count: %d\r\n", count);
        DBGUTxMsg(str);
    } while (count <= 10);

    while (1);
}
```

The loop consists of two keywords, **do** and **while**. Note also that the **while** statement in this loop is terminated with a semicolon, unlike in the normal **while** loop. The loop will always be entered into and the statements run at least once. The loop expression is then evaluated and if the result is true, the loop will be run again. **Figure 11.3** shows how the **do while** loop works.

When the **do_while** program is run, the count that is displayed is from 1 to 11 and not 1 to 10. This is because when a count of 10 is reached, the loop expression still evaluates to true. The loop is therefore run another time, incrementing the count to 11. When the loop expression is evaluated with a count value of 11, the result is false and the loop is exited.

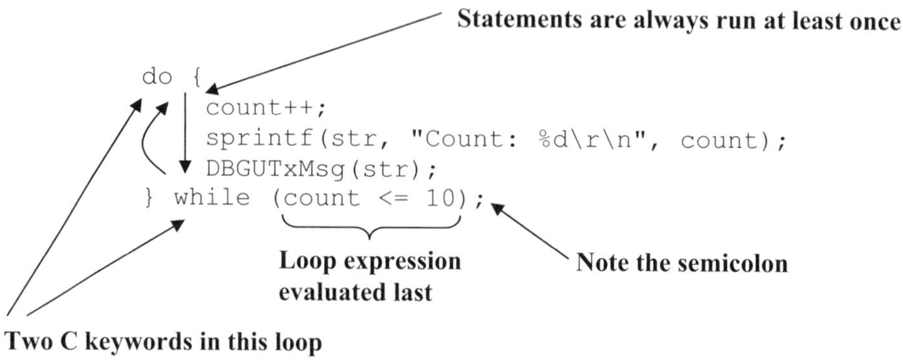

Figure 11.3: The `do while` Loop

11.4.3 The `for` Loop

The **`for`** loop is most often used in C programs for counting or doing something a certain number of times. The **`for_loop`** program that follows is similar to the **`water.c`** program of chapter four, except that it uses a **`for`** loop instead of a **`while`** loop.

```
#include <stdio.h>
#include "dbgu.h"

int main(void)
{
    char str[50];
    int level;

    DBGUInit(DBGU_115200);
    for (level = 0; level <= 10; level++) {
        sprintf(str, "Level = %d litres\r\n", level);
        DBGUTxMsg(str);
    }
    sprintf(str, "Finished");
    DBGUTxMsg(str);
    while (1);
}
```

Figure 11.4 illustrates how the **`for`** loop works.

```
        Initialisation     Control        Adjustment
for (level = 0; level <= 10; level++) {
        sprintf(str, "Level = %d litres\r\n", level);
        DBGUTxMsg(str);
    }
```

Figure 11.4: The `for` Loop

The **`for`** loop contains three expressions at the top of the loop body, namely, the initialisation expression, the control expression and the adjustment expression.

The initialisation expression is only evaluated once, when the **`for`** loop is first run in order to perform the necessary initialisation of the variable that will be used in the loop.

The control expression is evaluated every time that the loop is run, much like the loop expression in the **`while`** loop. If this expression evaluates to true, the statements in the body of the loop will be run. If it evaluates to false, the loop will be exited.

The adjustment expression is executed after each pass through the loop and before the control expression is evaluated again. This expression is used to change the value of the variable that will result in the loop being exited.

Here is how the **`for`** loop operates in the **`for_loop`** program:

1. Initialise the **`level`** variable to zero (**`level = 0`**)
2. Evaluate the control expression (**`level <= 10`**)
3. Run the statements in the loop body
4. Increment the **`level`** variable by one (**`level++`**)
5. Repeat steps 2 to 4 until step 2 evaluates to false, then exit the loop

11.5 Nested Loops and Decisions

Nesting in programming refers to putting one thing inside another. In the case of loops, we can put one loop inside another loop. With decisions, we can have a decision within a decision.

The **`nest_loop`** program shows one **`for`** loop inside another.

```
#include <stdio.h>
#include "dbgu.h"

int main(void)
{
    int count;
    char ch;
```

```
    char str[50];

    DBGUInit(DBGU_115200);
    for (count = 1; count <= 3; count++) {
        for (ch = 'A'; ch < 'E'; ch++) {
            sprintf(str, "%d, %c\r\n", count, ch);
            DBGUTxMsg(str);
        }
        sprintf(str, "----\r\n");
        DBGUTxMsg(str);
    }

    while (1);
}
```

The body of a **for** loop in this program contains another **for** loop. The second **for** loop is said to be nested in the main **for** loop. The main loop counts from 1 to 3. Every time that the loop executes the statements in the loops body, it runs the nested loop as well. The nested loop counts through the ASCII characters from A to D. Because the letters in the ASCII table are sequential, we can move from letter to letter by incrementing each character by one.

The nested loop could also use hexadecimal numbers or decimal numbers to achieve the same result. Each of the following **for** loops produces exactly the same results in this program:

```
for (ch = 'A'; ch < 'E'; ch++)
for (ch = 0x41; ch < 0x45; ch++)
for (ch = 65; ch < 69; ch++)
```

Every time that the main loop is run, the nested loop is run to completion. For every increment of 1 in the main loop, the nested loop counts through the characters A to D. After the nested loop is run, four "-" signs are printed out when the statements below the nested loop are run.

It is possible to nest any kind of loop into any other kind of loop. You can nest a **for** loop in a **while** loop, a **while** loop in a **for** loop, etc.

Decisions can also be nested. You have already seen a nested decision in the **serial_rx** program in chapter 10. Another example is shown in the **nest_if** program below.

```
#include <stdio.h>
#include "dbgu.h"

int main(void)
{
    char rx_data;
    char str[50];

    DBGUInit(DBGU_115200);
```

```
    while (1) {
        if (DBGUCheckForRxData()) {
            rx_data = DBGURx();
            if (rx_data == '1') {
                sprintf(str, "Menu item 1 selected\r\n");
            }
            else if (rx_data == '2') {
                sprintf(str, "Menu item 2 selected\r\n");
            }
            else {
                sprintf(str, "%c invalid selection\r\n", rx_data);
            }
            DBGUTxMsg(str);
        }
    }
}
```

This program continually checks if a character has been received on the serial port. Use HyperTerminal to send characters on the serial port. When a character has been received, it is first read from the serial port and stored in the **rx_data** variable. A nested **if** statement is used to check if the received character is a valid menu option. The use of indenting helps to show which **else** belongs to which **if** statement (**figure 11.5**).

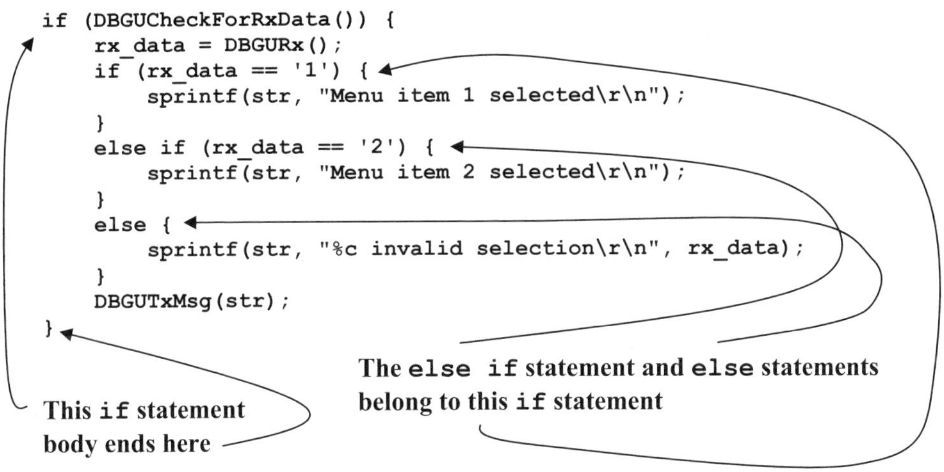

Figure 11.5: Nested Decisions

11.6 Decision Making with the `switch` Statement

The **menu** program will demonstrate the use of the **switch** statement which is similar to using **else if** in C.

```
#include "dbgu.h"

int main(void)
```

```
{
    char rx_data;

    DBGUInit(DBGU_115200);
    while (1) {
        while (!DBGUCheckForRxData());
        rx_data = DBGURx();
        switch (rx_data) {
            case '1':
                DBGUTxMsg("Menu item 1.\r\n");
            break;

            case '2':
                DBGUTxMsg("Menu item 2.\r\n");
            break;

            case '3':
                DBGUTxMsg("Menu item 3.\r\n");
            break;

            case '4':
                DBGUTxMsg("Menu item 4.\r\n");
            break;

            default:
                DBGUTxMsg("Invalid selection.\r\n");
            break;
        }
    }
}
```

When this program is run, pressing keys 1 to 4 in HyperTerminal will select various menu items. If anything else is pressed, a default message is displayed to say that the selection was invalid.

Four C keywords are associated with the **switch** construct, namely **switch**, **case**, **default** and **break**. The **switch** statement uses a set of parentheses that contain a variable or expression that is to be selected for in the body of the switch statement. Inside the body of the switch statement are a number of **case** statements. The **switch** variable or expression is checked to see if it matches one of the **case** statements. If it does, the statements under the **case** statement are run in the order that they appear. To stop running any further unwanted statements, the **break** keyword is used to break out of the body of the **switch** statement. If **break** was not used, all of the other statements below the **case** would run. The statements under the **default** keyword will be run if no match is found at any **case** statement. The **switch** statement does the same as the following code, but using **switch** is easier to read and more flexible:

```
if (rx_data == '1') {
    DBGUTxMsg("Menu item 1.\r\n");
}
else if (rx_data == '2') {
    DBGUTxMsg("Menu item 2.\r\n");
```

```
}
else if (rx_data == '3') {
    DBGUTxMsg("Menu item 3.\r\n");
}
else if (rx_data == '4') {
    DBGUTxMsg("Menu item 4.\r\n");
}
else {
    DBGUTxMsg("Invalid selection.\r\n");
}
```

The **`default`** statement can be left out if desired and is the equivalent of leaving off the **`else`** statement in the code example above. **Figure 11.6** shows the details of the switch statement. In the figure, only the case of **`rx_data`** containing a value of 1 is shown, but each of the other cases will behave the same way – e.g. if **`rx_data`** contains a value of 2, the code under **`case '2'`** will run and then the **`break`** statement will exit the switch.

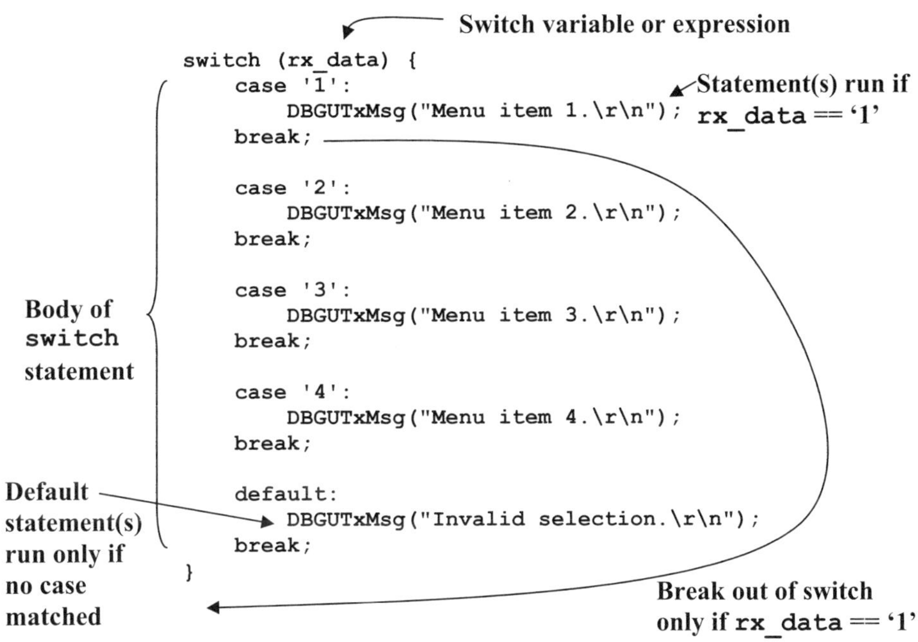

Figure 11.6: The switch Statement

11.7 The Conditional Operator

Another operator used for making decisions is known as the conditional operator. The **`sw_condition`** program demonstrates its use.

```
#include <stdio.h>
#include "at91sam7s.h"
#include "dbgu.h"

int main(void)
{
    unsigned int sw_state = 0;
    char str[50];
    /* enable the clock of the PIO controller */
    PMC_PCER = 0x00000004;
    /* configure PIO for switch */
    PIO_PER = 0x00008000;
    PIO_ODR = 0x00008000;           /* pins used as input */
    DBGUInit(DBGU_115200);
    while (1) {
        sw_state = (PIO_PDSR & 0x00008000) ? 1 : 2;
        sprintf(str, "Switch state: %d\r", sw_state);
        DBGUTxMsg(str);
    }
}
```

This program uses one of the push button switches on the embedded system, so make the necessary changes to your program so that it uses one of the push buttons interfaced to your board. The program reads the switch attached to microcontroller pin PA15.

When run, the program monitors the push button switch and will send a message containing either 1 or 2, depending if the switch is pushed or not, on the serial port to HyperTerminal. The conditional operator is used to choose the number to send, depending on the state of the switch. **Figure 11.7** shows how the conditional operator works.

The conditional operator consists of a *condition* that will evaluate to either true or false as we have previously seen in C. Should the result of the *condition* be true, the first *expression* will be the result of the operation. In the example program, the **`sw_state`** variable will be assigned the value of 1 when the *condition* is true. I.e. in this program when the switch is pressed, as the state of the switch is the condition that is being checked.

Should the *condition* evaluate to false, the second *expression* will be the result and will therefore be substituted into the **`sw_state`** variable in the example. The **`sw_state`** variable will then contain the value 2 in this example.

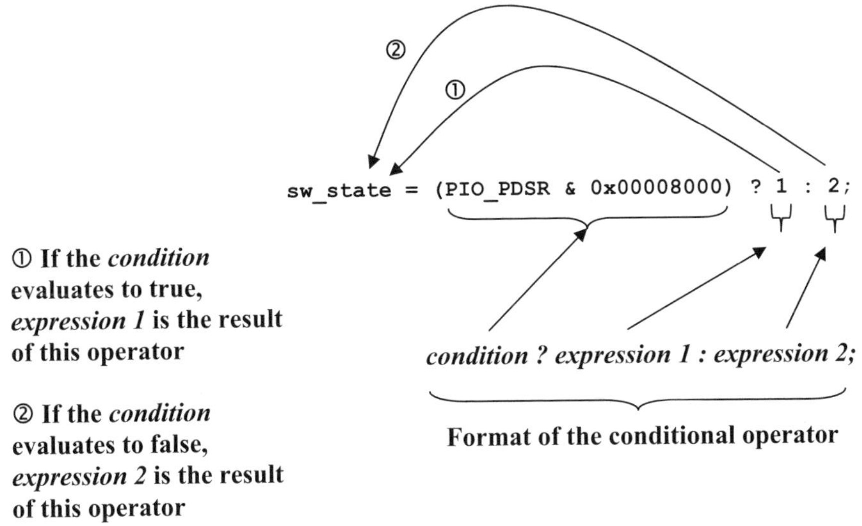

Figure 11.7: The Conditional Operator

11.8 Functions and Pointers

When you pass a variable to a function, the function is sent a copy of the variable and operates on this local copy of the variable. If the function modifies the variable, the original variable that was passed to it remains unmodified.

It is possible in C to pass the address of a variable to a function. The function will then be able to modify the value of the variable as it has the address of the variable and not a copy of the variables contents. The next program, **point_func**, shows how this works.

```
#include <stdio.h>
#include "dbgu.h"

void func1(int num);
void func2(int *pnum);

int main(void)
{
    char str[50];
    int number = 123;

    DBGUInit(DBGU_115200);
    sprintf(str, "Number is %d\r\n", number);
    DBGUTxMsg(str);
    func1(number);
    sprintf(str, "Number is %d\r\n", number);
    DBGUTxMsg(str);
    func2(&number);
    sprintf(str, "Number is %d\r\n", number);
```

```
    DBGUTxMsg(str);
    while (1);
}

void func1(int num)
{
    char str[50];
    num = 567;
    sprintf(str, "Func1 num %d\r\n", num);
    DBGUTxMsg(str);
}

void func2(int *pnum)
{
    char str[50];
    *pnum = 987;
    sprintf(str, "Func2 pnum %d\r\n", *pnum);
    DBGUTxMsg(str);
}
```

When this program is run, the following output is produced and displayed on HyperTerminal:

Number is 123
Func1 num 567
Number is 123
Func2 pnum 987
Number is 987

The variable **number** is first printed out and displays the value that it was initialised to (**123**). This variable is then passed to the function **func1()**. This function changes the value of the variable and prints the result out (**567**). When the **number** variable was passed to **func1()**, a copy of the variables contents was made to a "local" variable in **func1()** called **num**. This local variable was modified to contain the number 567 and left the variable that was passed to it unaffected. This is confirmed when after returning from **func1()**, the value of **number** is printed out again and remains unchanged (**123**).

The function **func2()** is passed the address of the variable **number** (**&number**). This address is stored in the **pnum** pointer variable which is local to **func2()**. Inside this function, the address variable is dereferenced (***pnum**) in order to operate on the contents of the variable that it is pointing to. It is of course pointing to the variable **number** and not to a local copy of **number**. The line of C code in **func2()**:
***pnum = 987;**
means "assign the value of 987 to the memory location that **pnum** is pointing to". **pnum** contains the address of **number** and is therefore pointing to it. The value of the variable that **pnum** is pointing to is then printed out inside **func2()** (**987**).

After returning from **func2()**, the value of **number** is then printed out again (**987**). It is now 987 and not 123 anymore. This is because we modified the contents of this variable by passing it's address to **func2()**.

11.8.1 Returning More Than One Value from a Function

Pointers enable us to return more than one value from a function. The **return2** program uses two pointers to return two values from a function.

```
#include <stdio.h>
#include "dbgu.h"

void retfunc(int *val1, int *val2, int input);

int main(void)
{
    char str[50];
    int num1, num2;

    DBGUInit(DBGU_115200);
    retfunc (&num1, &num2, 34);
    sprintf(str, "34 squared = %d\r\n", num1);
    DBGUTxMsg(str);
    sprintf(str, "34 * 10 = %d\r\n", num2);
    DBGUTxMsg(str);
    while (1);
}

void retfunc(int *val1, int *val2, int input)
{
    *val1 = input * input;
    *val2 = input * 10;
}
```

The **retfunc()** function in this program takes two pointer arguments and one integer argument. The pointer arguments are used to store the two output values from the function into the two variables that they point to.

When **retfunc()** is called, it is passed the address of the variable **num1** and the address of variable **num2**. The constant 34 is also passed to the function in the function variable **input**. The function then calculates the square of the variable **input** (**input** multiplied by itself) and saves it in the variable that pointer **val1** is pointing to i.e. **num1**.

Next, **retfunc()** calculates **input** multiplied by 10 and stores it in the variable that pointer **val2** is pointing to, i.e. **num2**. The function has thus returned two values which it would not normally be able to do. After returning to **main()**, the calculated values are found in the variables **num1** and **num2** and are printed out.

11.9 Variables and Scope

If you look at the **point_func** program, you will notice each function, including **main()** needed it's own **str** array. Also notice that the array used the exact same name i.e. **str** in each function without the compiler complaining. This is because variables defined in a function have what is known as "function scope". A variable defined in **main()** can only be used and "seen" inside **main()**. A function can only "see" the variables defined inside it and does not know about variables that are defined inside any other function.

If you try to use a variable inside a function that was defined in another function, the compiler will issue an error. It is however possible to define a new variable inside a function with the same name as one that has already been defined in another function. Variables that are defined inside a function are known as "local" variables as they are local to the function.

11.9.1 Global Variables

Global variables have "file scope", meaning that they can be "seen" and used by every function in a file – the C source file that they are defined in. A global variable is also known as an "external" variable. The **glob_var** program uses a global variable to pass information between functions.

```
#include <stdio.h>
#include "dbgu.h"

int glob_count = 0;

void AddOne(void);
void Delay(void);

int main(void)
{
    char str[50];

    DBGUInit(DBGU_115200);
    while (1) {
        sprintf(str, "Count is %d\r", glob_count);
        DBGUTxMsg(str);
        AddOne();
        Delay();
    }
}

void AddOne(void)
{
    glob_count++;
}

void Delay(void)
{
```

```
    volatile int del = 1000000;
    while (del--);
}
```

In this program, the variable **glob_count** is defined as a global variable by defining it outside of **main()**. All functions inside the file in which it exists have access to it. In the program, **glob_count** is incremented in the **AddOne()** function and displayed in the **main()** function. Both these functions have therefore accessed this global variable.

Global variables can be helpful, but it is not considered good programming practice to use too many as one can loose track of which functions are changing these variables in a big program.

11.10 Static Variables

Variables that are declared within a function, i.e. local variables, are destroyed when the function returns. The memory that they use when the function is running is freed up for use by other parts of the program when they return. The **static** keyword can be used to prevent local variables from being destroyed as the **static_var** program demonstrates.

```
#include <stdio.h>
#include "dbgu.h"

int Count(void);
int StaticCount(void);
void Delay(void);

int main(void)
{
    char str[50];
    DBGUInit(DBGU_115200);

    while (1) {
        sprintf(str, "count %d, static count, %d\r",
                Count(), StaticCount());
        DBGUTxMsg(str);
        Delay();
    }
}

int Count(void)
{
    int count = 0;
    return count++;
}

int StaticCount(void)
{
    static int count = 0;
    return count++;
}
```

The **`Delay()`** function has been left out of the listing above, but is the same as the previous program. This program calls two functions that perform identical actions of incrementing a local variable and returning the result. The only difference between the functions is that one uses a static variable and the other does not. Note that the post increment operator is used in these functions, so the count is first returned and then incremented.

The non-static variable is always initialised to 0 when the function is called and it effectively "forgets" the value of this variable between calls. The result is that when the returned value is printed, a 0 is always displayed.

The **`StaticCount()`** function uses a static local variable. The variable is initialised to 0 the first time that the function is called and thereafter "remembers" the value that the variable contains between function calls. The result is that every time this function is called in the program, the next sequential count value is returned.

11.11 Floating Point Data Types

We covered the integer data types in chapter 7, we will now take a look at the floating point data types. Floating point numbers are used to store numbers that have a decimal point. **Table 11.3** shows the three floating point data types in C.

Type	Size	Range	Precision
`float`	4 byte	±3.4E+38	6 digits
`double`	8 byte	±1.7E+308	15 digits
`long double`	10 bytes	±1.1E+4932	19 digits

Table 11.3 Floating Point Data Types

You have already used the **`float`** type, the **`double`** and **`long double`** work the same way, but have higher precision. Precision refers to the number of digits that the number can consist of.

11.12 Casts

Casting was seen in chapter 8 when we created a pointer to a memory mapped register e.g:

```
(*((volatile unsigned int *) 0xFFFFF400))
```

A cast has the format: (*type name*) *expression*
This forces the *expression* to become the type represented by *type name*. The following example casts an **`int`** type to a **`char`** type and a **`float`** type to an **`int`** type.

```
int an_int = 4567;
char ch;
float flt = 1.234;

ch = (char)an_int;
an_int = (int)flt;
```

The first cast tells the compiler that you want to convert an **int** to a **char**. Because a cast is used, the compiler will not complain. The compiler would normally complain because a **char** is only one byte long and an **int** is four bytes long. The resulting operation will put the least significant byte (lowest 8 bits) of the **int** into the **char**, the upper 24 bits will be discarded.

The second cast forces a floating point number to become an integer. This will result in only the integer part of the floating point number being assigned to the integer and the fractional part after the decimal point being discarded.

11.13 Exercises

1. Use a **for** loop to count down from 10 to 0 and send each count value to be displayed on HyperTerminal. When counting down, use the initialisation expression to initialise the loop variable to 10, the control expression to break out of the loop when the loop variable is less than 0 and the adjustment expression to decrement instead of increment the loop variable.

2. Use a **switch** statement that has four cases to switch two LEDs individually on or off by a user sending characters from a keyboard. When ‘a’ is sent, switch LED1 on, when ‘b’ is sent, switch LED1 off, when ‘c’ is sent, switch LED2 on and when ‘d’ is sent, switch LED2 off.

11.13.1 Solutions

1. **count_down**:

```
#include <stdio.h>
#include "dbgu.h"

int main(void)
{
    char str[50];
    int count;

    DBGUInit(DBGU_115200);

    for (count = 10; count >= 0; count--) {
        sprintf(str, "count = %d\r\n", count);
        DBGUTxMsg(str);
    }
    while (1);
```

```
}
```

2. **`LED_select`**:

```
#include "at91sam7s.h"
#include "dbgu.h"

int main(void)
{
    char rx_byte;

    /* using LEDs on pins PA0 and PA1 */
    PIO_PER = 0x00000003;        /* enable uC pins as PIO */
    PIO_OER = 0x00000003;        /* enable uC pins as output */
    DBGUInit(DBGU_115200);

    while (1) {
        if (DBGUCheckForRxData()) {      /* is a byte received? */
            rx_byte = DBGURx();          /* yes, get the byte */

            switch (rx_byte) {
                case 'a':
                    PIO_CODR = 0x00000001;  /* switch LED1 on */
                break;

                case 'b':
                    PIO_SODR = 0x00000001;  /* switch LED1 off */
                break;

                case 'c':
                    PIO_CODR = 0x00000002;  /* switch LED2 on */
                break;

                case 'd':
                    PIO_SODR = 0x00000002;  /* switch LED2 off */
                break;
            }
        }
    }
}
```

11.14 Summary

- Format specifiers and escape sequences work with **`sprintf()`** as with **`printf()`**.
- The **`break`** and **`continue`** statements are used to change the flow of a program and are commonly used in loops.
- The **`do while`** loop will always run the body of the loop once and then check the loop expression.

- The **`for`** loop is most often used to count or perform some action a set number of times.

- Loops and decisions can be nested in C. Using proper indenting in source code makes nesting easier to read and understand so be sure to always indent a program clearly.

- The **`switch`** statement is another decision making construct in C which can be easier to read than using the **`if`** – **`else if`** – **`else`** construct especially when there are many comparisons being made.

- The conditional operator is another decision making operator in C.

- More than one value can be returned from a function by passing the addresses of variables to a function so that the function can store the values to be returned in these variables.

- All variables in a C program have "scope". The scope of a variable determines which functions in the program can "see" and use the variable. Variables defined within a function have function scope as they can only be used by the function in which they are defined. Global variables have file scope and can be seen by all of the functions within the file that they are defined.

- Static variables are not "destroyed" (freed up) when a function returns. A static variable and its contents are preserved between function calls to the function that contains the static variable.

- Casts allow a programmer to force one type of variable's value to be stored in another type of variable.

12 Arrays and Strings

What you will do in this chapter:

- See how arrays are used to access sequential memory locations
- See how strings are used and stored in memory
- Use C standard library string functions
- Write functions that take strings and arrays as arguments

You have already worked with strings in C when using the **`printf()`** and **`sprintf()`** functions. C does not have a string data type but instead uses arrays of characters. An array of characters is simply a number of sequential characters in memory. So in C, a string is an array of characters, but it is also possible to have and array of a different data type. We will start by looking at arrays in general and see how to define and use arrays in C. We will then take a closer look at strings and some standard library functions used to manipulate them.

12.1 Arrays

An array is a set of consecutive memory locations that store data of a specified type. The keyword here is *consecutive* as this is what makes arrays possible. As an example, in C an array of 5 integers is defined as follows:

```
int arr_num[5];
```

This tells the compiler to make space for an array that consists of five integer numbers. Each element (integer) in the array can be accessed using array notation in C. If we wanted to assign a value to each element in the array, it can be done as follows:

```
arr_num[0] = 20;
arr_num[1] = 34;
arr_num[2] = 9;
arr_num[3] = 1023;
arr_num[4] = 700;
```

There are five elements in this array, but they are indexed starting at 0, so the five elements are accessed with indexes from 0 to 4 and not 1 to 5.

The array definition:

```
int arr_num[5];
```

is similar to defining 5 separate integer variables:

```
int arr_num0, arr_num1, arr_num2, arr_num3, arr_num4;
```

Defining separate variables, however does not guarantee that they are consecutively placed in memory. The separate variables can also not be referenced with array notation which, as we will see, is very useful.

The **index_array** program will initialise an array of integers and then use a **for** loop to move through the list of integers and send them out of the serial port.

```
#include <stdio.h>
#include "dbgu.h"

int main(void)
{
    int num_list[5] = {34, 789, 327, 1000, 45};
    char str[50];
    int index;

    DBGUInit(DBGU_115200);
    for (index = 0; index < 5; index++) {
        sprintf(str, "Element %d = %d\r\n", index, num_list[index]);
        DBGUTxMsg(str);
    }
    while (1);
}
```

The first line inside **main()** defines an array of 5 integers and initialises each of these integers to a value that appears between braces after the = sign. The initialisation has the same effect as assigning a value to each element of the array as follows:

```
int num_list[5];
num_list[0] = 34;
num_list[1] = 789;
num_list[2] = 327;
num_list[3] = 1000;
num_list[4] = 45;
```

A **for** loop is used to count from 0 to 4. Inside this loop, the current count value (the variable **index**) and each element of the array is printed out. By referencing each element in the array with the current count that is counting from 0 through 4, we access each element in the array from 0 to 4.

When **index** contains the value 0, then **num_list[index]** takes on the value of the first element in the array i.e. 34 in this example. After **index** has been incremented to 1, **num_list[index]**, which is equivalent to **num_list[1]**, now has the value of the second element in the array i.e. 789. Remember that we are counting from 0, so the first element in the array is indexed at position 0, the second at position 1 the third at position

2, etc. **Figure 12.1** shows the array used in this example in diagrammatic format as it would appear in memory.

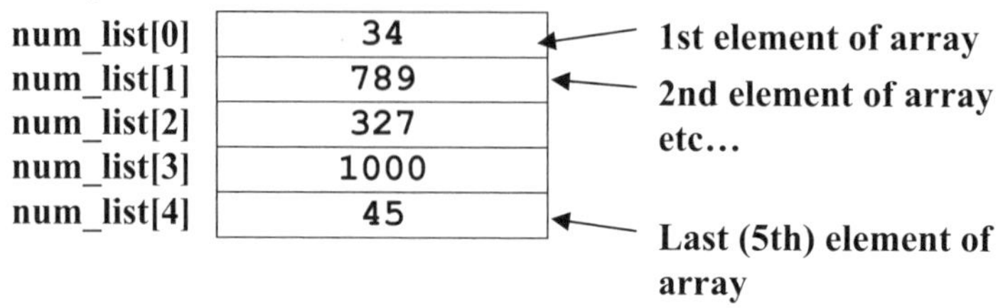

Figure 12.1: An Array of Integers in Memory

The next program called **LED_seq** (LED sequence) initialises an array with a series of numbers that are written to the LEDs to make them move through a sequence of different combinations of off and on LEDs. An array used in this way is also known as a lookup table. The program sequences four LEDs on port pins PA0 to PA3, so make the necessary changes to initialise the LEDs on your board and change the values in the array to switch the correct LEDs on and off. The values in the array are the values that are written to the microcontroller port to switch the LEDs on and off. This program does not make use of the serial link and HyperTerminal, just 4 LEDs.

```
#include "at91sam7s.h"

void Delay(void);

int main(void)
{
    unsigned int seq[] = {0x00000001, 0x00000002, 0x00000004,
                          0x00000008, 0x00000006, 0x00000009,
                          0x0000000F, 0x00000005, 0x0000000A,
                          0x00000005, 0x0000000A};
    int ind;
    /* configure PIO lines for LEDs on PA0 to PA3 */
    PIO_PER = 0x0000000F;       /* enable uC pin as PIO */
    PIO_OER = 0x0000000F;       /* enable uC pin as output */
    PIO_SODR = 0x0000000F;      /* switch all LEDs off */

    while (1) {
        for (ind = 0; ind < (sizeof(seq) / 4); ind++) {
            PIO_CODR = seq[ind];
            Delay();
            PIO_SODR = seq[ind];
        }
    }
}

void Delay(void)
{
    int del = 5000000;

    while (del--);
}
```

An array of type **`unsigned int`** is defined at the beginning of **`main()`** that contains the on and off sequences of the LEDs. **Figure 12.2** shows what the numbers in the array produces when written to the LEDs.

Number	LED Pattern
0x00000001	□ □ □ ■
0x00000002	□ □ ■ □
0x00000004	□ ■ □ □
0x00000008	■ □ □ □
0x00000006	□ ■ ■ □
0x00000009	■ □ □ ■
0x0000000F	■ ■ ■ ■
0x00000005	□ ■ □ ■
0x0000000A	■ □ ■ □
0x00000005	□ ■ □ ■
0x0000000A	■ □ ■ □

■ = LED ON
□ = LED OFF

Figure 12.2 : The Sequence of LED Patterns in the `LED_seq` Program

In this program, an array of unsigned integers is defined and each one initialised to a value that will be written to the PIO port. The notation for defining the array is:

```
unsigned int seq[]
```

The number of elements in this array is missing, and would normally be placed between the square brackets as in the previous example. Because we have initialised the array with 11 values, the compiler will count the values and make the array the correct size of 11. If we did not initialise the array, we would have to specify the size of the array, as the compiler would then not know how big we intended to make the array.

The PIO port is then setup to drive four LEDs on pins PA0 to PA3. The LEDs are all switched off before the **`while(1)`** loop is entered.

Inside the **`while(1)`** loop is a **`for`** loop that is used to access each element of the array in sequence. When the **`for`** loop has counted to the last element of the array, it will start at the beginning of the array again. It counts through the 11 elements of the array by counting from 0 to 10 continually. The number to count up to is obtained by using the C **`sizeof`** keyword. This keyword returns the size of the array in bytes. Because each element of the array is in words which are 4 bytes long on the ARM, the result is divided by 4 to convert bytes to words. The reason for calculating the maximum value to count to in the **`for`** loop this ways is to make it easier to change the patterns that are written out to the LEDs. To change the number of patterns, simply add or remove more values where the array is initialised. The size of the array is automatically calculated by the compiler and the maximum number that the **for** loop must count to is also adjusted automatically.

Inside the **for** loop, each number in the array (LED pattern) is written to the PIO port. A delay function is then called to leave the written LED pattern on for a while. The LEDs are then all switched off and the next pattern immediately written to the LEDs so that they appear to have just changed patterns and not switched off at all.

We have so far seen only arrays of integers, but it is possible to have arrays of other C data types.

An array of floating point numbers:

```
float sample[30];
```

An array of characters:

```
char chr_array[10];
```

We will now look at strings which are just arrays of characters.

12.2 Strings

Strings are useful in embedded systems wherever status information is to be displayed to a user or received from a keypad or keyboard input. Strings can be displayed on an LCD display or on a terminal emulator as we have been doing or even saved to SD card or disk if the embedded system contains one.

Strings are sequences or arrays of characters that are NULL terminated. This means that the last character in a string must have the value of 0. The NULL is the first number in the ASCII table and has the escape sequence **\0**.

You have already used string constants and arrays for storing strings. The **strings** program will demonstrate the use of these string constants and string variables. You will need to connect to HyperTerminal again to run this program.

```
#include <stdio.h>
#include "dbgu.h"

int main(void)
{
    char str[50];
    char msg[4];

    DBGUInit(DBGU_115200);
    sprintf(str, "%s\r\n", "Hello");
    DBGUTxMsg(str);
```

```
    msg[0] = 'B';
    msg[1] = 'y';
    msg[2] = 'e';
    msg[3] = 0;
    DBGUTxMsg(msg);
    while (1);
}
```

In this program the familiar character array **str** is used again for writing strings to, using **sprintf()**. An additional character array, **msg**, is also defined and is allocated 4 bytes of memory.

An example of a string constant is the string **"Hello"**, used in the **sprintf()** statement. This string cannot be changed after compiling when the program is running and is therefore constant.

The **msg** array is used to demonstrate how a string is built up of characters and terminated with 0. It is known as a string variable as its contents can be modified when the program is running. As can be seen, the word "Bye" is written to the **msg** array by writing one character at a time to each element of the array. A zero is written to the last element in the array in order to correctly terminate the string which is now stored in this array. The escape sequence **\0** could also be used to terminate the string and would also result in a zero being written to the last element in the array, it would need to be written between single quotes i.e. **'\0'**. The size of an array used to store a string must always be one element longer than the string because of the NULL terminating character of the string. We could have made the **msg** array in this program longer than 4 elements; the other elements of the array would just not be used. If the array was made too short and we tried to write a character to the array that was beyond the end of the array, the compiler would complain. **Figure 12.3** shows what the string constant and string variable from the **strings** program look like in memory. The addresses shown would be whatever addresses were assigned to the strings by the linker. It is not important what these addresses are; only that they fall in the correct range of the microcontroller's memory map and that each element in the string is consecutively placed in memory (consecutive addresses).

The string variable **str** is not shown in this figure. In the program, **sprintf()** is used to copy the string constant **"Hello"** to the string variable **str**. The string variable **str** was made extra long i.e. 50 bytes, in order to prevent any function that uses it from writing to memory beyond this array. In C there is no bounds checking of arrays which makes it possible for a program to write data beyond an array and modify data that it should not. It is up to the programmer to do bounds checking or to make an array sufficiently large enough so that a program will not be able to write to memory that is beyond the array.

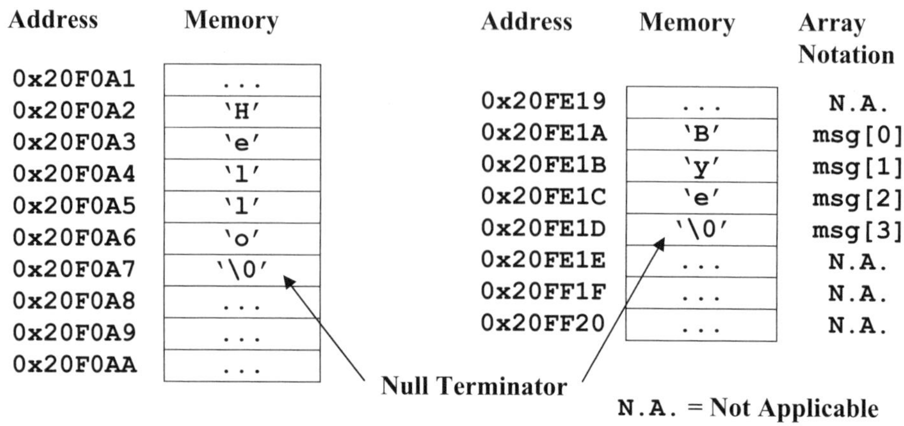

Figure 12.3: The "String Constant" (left) and "String Variable" (right) used in the `string` Program

12.2.1 Writing to a String

The **`write_string`** program allows a user to send a string to the embedded system via HyperTerminal.

```
#include <stdio.h>
#include "dbgu.h"

#define MAX_STR_LEN     35

int main(void)
{
    char str[50];
    char in_str[MAX_STR_LEN];
    int index = 0;

    DBGUInit(DBGU_115200);
    sprintf(str, "\r\nEnter a string, maximum length %d\r\n",
                                                MAX_STR_LEN - 1);
    DBGUTxMsg(str);
    while (index < (MAX_STR_LEN - 1)) {
        if (DBGUCheckForRxData()) {
            in_str[index] = DBGURx();
            if (in_str[index] == '\r') {
                break;
            }
            DBGUTx(in_str[index]);
            index++;
        }
    }
    in_str[index] = 0;
    DBGUTxMsg("\r\nYou entered the string:\r\n");
    DBGUTxMsg(in_str);
    while (1);
}
```

This program sends a message to the user on HyperTerminal and asks him to enter a string. As the user enters each character of the string, each character is sent back to HyperTerminal (echoed) so that the user can see what he is typing. Note that if you have local echo enabled on HyperTerminal, you will see every letter printed twice.

When the user presses the **Enter** key, the program null terminates the string and then displays the string on HyperTerminal. If the user types more characters than can fit in the string variable array, the string is also null terminated and displayed before the user can press the **Enter** key.

Let's have a look at how the **write_string** program works. Firstly a maximum size for the string that the user will enter is defined using the **#define** directive. If you would like to change the maximum size of string that the user can enter in this program, you just need to change it in one place, where it is defined, and the size and bounds checking throughout the program will automatically be adjusted.

A message is sent to HyperTerminal, prompting the user to enter a string. The maximum length of the string that can be entered by the user is also sent. This value is calculated by subtracting 1 from the length of the array that is used to store the string that will be input by the user. The reason for subtracting 1 from this length is because the null terminating character needs to be stored in the last element of the array if the maximum number of characters allowed in the string is used.

A **while** loop is used to get the characters from the user via the serial port. The **while** loop is set to terminate if the user types more characters than can be stored in the array set aside for these characters. This is done by checking the **index** variable that is used to access each element of the array to see if it has gone out of bounds. We actually have to check this value against the length of the input array minus 2. This is firstly to accommodate the null termination character and secondly because the last element in the array is one less than the length of the array (because the first element in the array is counted as 0 and not as 1). The loop expression is as follows:

```
index < (MAX_STR_LEN - 1)
```

This checks that the index is less than the maximum string length minus one, which is equivalent to checking that the index is less than or equal to the maximum string length minus two. In other words we could have written:

```
index <= (MAX_STR_LEN - 2)
```

Inside the **while** loop, there is an **if** statement that checks to see if a character was received on the serial port. The body of the **if** statement is only run if a character is received.

When a character is received, it is saved to the character array **in_str**. The variable **index** is used to determine which element of the array the character is saved to. This variable is initialised to 0, so stores the first character to the first element (element 0) of the array. The **index** variable is incremented after each character is saved so that the next character will be saved to the next element in the array.

After a character has been saved to the array, the character is checked to see if it is the carriage return character (**\r**) i.e. we are checking to see if the user pressed the **Enter** key. If the **Enter** key was pressed, the **break** statement is used to exit out of the **while** loop without executing a further statement in the **while** loop. If **Enter** was not pressed, the character that the user typed is sent back to HyperTerminal via the serial port by calling the **DBGUTx()** function which transmits a single character. This function is one of the driver functions for the DBGU port from the project template files and is found in the **dbgu.c** source file. After echoing the typed character back to the user, the **index** variable is incremented so that it will access the next element in the **in_str** array. The **while** loop's loop expression is checked to see if **index** is out of bounds or not and then the **if** statement in the loop is run again to wait for the next character from the serial port.

The **while** loop is exited if the user enters too many characters or if the **Enter** key is pressed. If either of these two conditions occur, the null terminator is added to the string in the element that **index** is accessing. This would be the last element in the array if the loop was exited because the array was full. In the case of **Enter** being pressed, the **\r** character is saved to the array and then overwritten by the null terminating character as **index** is not incremented after breaking out of the loop.

A text message is then displayed on HyperTerminal and finally the string that the user entered, which is now saved in the **in_str** array, is also sent.

12.2.2 Initialising a String

The **init_string** program shows two ways of initialising a string.

```
#include "dbgu.h"

int main(void)
{
    char str1[] = { 'F', 'i', 'r', 's', 't', '\r', '\n', '\0' };
    char str2[] = "Second\r\n";

    DBGUInit(DBGU_115200);
    DBGUTxMsg(str1);
    DBGUTxMsg(str2);
    while (1);
}
```

Because a string is an array, we can initialise a string in the same way that we initialised an array of integers. The string **str1** in this example is initialised with individual characters. Note that the null termination must also be put into the string.

The second string, **str2**, in this program shows an easier way of initialising a string by putting it between double quotes. When this method is used, the compiler automatically inserts the null terminating character at the end of the string.

12.2.3 C Library String Functions

The C standard library provides a number of functions that are used for manipulating strings. The prototypes for these functions are found in the header file **string.h**. **Table 12.1** shows some of these string functions. The **str_func** program demonstrates the use of these functions.

Function	Description
strcat(*str1*, *str2*)	Concatenates (appends) *str2* to *str1*
strcmp(*str1*, *str2*)	Compares two strings. Returns 0 if the strings are equal, < 0 if *str1* is less than *str2*, > 0 if *str1* is greater than *str2*
strcpy(*str1*, *str2*)	Copies *str2* into *str1*
strlen(*str*)	Returns the length of the string *str*, excluding the null-terminating character

Table 12.1 Some of the String Functions Available in the C Standard Library

```
#include <stdio.h>
#include <string.h>
#include "dbgu.h"

int main(void)
{
    char str1[10] = "abc";
    char str2[] = "def";
    char str[50];

    DBGUInit(DBGU_115200);
    sprintf(str, "\fThe length of str1 is %d\r\n", (int)strlen(str1));
    DBGUTxMsg(str);
    sprintf(str, "Comparing str1 & str2: %d\r\n", strcmp(str1, str2));
    DBGUTxMsg(str);
    sprintf(str, "Comparing str2 & str1: %d\r\n", strcmp(str2, str1));
    DBGUTxMsg(str);
    strcat(str1, str2);
    sprintf(str, "str1 is now: %s\r\n", str1);
    DBGUTxMsg(str);
    strcpy(str1, str2);
    sprintf(str, "str1 is now: %s\r\n", str1);
    DBGUTxMsg(str);
    sprintf(str, "Comparing str1 & str2: %d\r\n", strcmp(str1, str2));
    DBGUTxMsg(str);
    while (1);
}
```

The string **str1** is forced to make space for a string that is 9 characters long plus the null-terminating character by specifying the length of the string as 10 between the square brackets when the string is defined. **str1** is however initialised with a string that is only 3 characters long. The reason for allocating more space than necessary is so that there will be enough space to concatenate a string to this one without going out of the array bounds.

The **strlen()** functions is called first and returns the length of string **str1**. The length returned is 3 which is the length of the string excluding the null-terminating character. The return value of **strlen()** is cast to an **int** in order to prevent the compiler from issuing a warning as the return type of this function is a specially defined type.

strcmp() is called twice to compare the two initialised strings. The comparison is done by this function a character at a time. The difference between the first set of characters that do not match is returned. In the example, “abc” (**str1**) is compared with “def” (**str2**). The first characters that do not match are ‘a’ and ‘d’. Looking at the ASCII table, one can see that these characters have values of 97 and 100. The difference between the two values is 100 – 97 = 3. When these strings are compared again, they are compared in reverse order, so the difference between the two is now 97 – 100 = -3.

The **strcat()** function is called to concatenate **str2** onto **str1**. Some of the extra space made available when defining **str1** is now used to store the extra string added to the end of the string contained in **str1**. **str1** changes from “abc” to “abcdef” after concatenation.

strcpy() is called to copy **str2** into **str1**. This action overwrites the contents of **str1** and replaces it with the contents of **str2**.

strcmp() is called again to compare the two strings. Because **str2** was copied to **str1**, they are now equal and **strcmp()** returns a value of 0.

12.3 Arrays and Addresses

The name of an array is the address of the array. For example if we have an array of characters as follows:

```
char processor[] = "ARM";
```

The address of the array and therefore the address of the first element in the array is the name of the array in C and in this example is **processor**, we do not get the address by writing **&processor** as with single variables.

The value of the first element in the array is obtained by using array notation **processor[0]**. The contents of each of the other elements in the array can be accessed in a similar fashion e.g. **processor[1]**, etc. The address of any element can be obtained by using the address operator with array notation e.g. **&processor[1]** is the address of the second element in this array. The **array_addr** program will clarify.

```
#include <stdio.h>
#include "dbgu.h"

int main(void)
{
    char processor[] = "ARM";
    char str[50];
    int i;

    DBGUInit(DBGU_115200);
    sprintf(str, "\fAddress of array is %d\r\n", (int)processor);
    DBGUTxMsg(str);
    for (i = 0; i < (sizeof(processor) - 1); i++) {
        sprintf(str, "addr = %d, data = %d (%c)\r\n",
                (int)&processor[i], processor[i], processor[i]);
        DBGUTxMsg(str);
    }
    while (1);
}
```

This program takes the string "ARM" and prints out the address of the array (i.e. the address of the first element of the array). It then prints out the address of each element in the array as well as its contents in decimal and character format.

The output of this program is:

```
Address of array is 2162248
addr = 2162248, data = 65 (A)
addr = 2162249, data = 82 (R)
addr = 2162250, data = 77 (M)
```

When the address of the array is printed, it is cast to an integer (**(int)processor**) to tell the compiler that we want to display the address of the array as an integer. The address of the array is a pointer to a character (**char ***) so the cast stops the compiler complaining as we specify that we want to print this address as an integer.

The **for** loop in this program prints out the address of each element of the array as well as the contents of each element of the array as previously explained. The terminating condition of the **for** loop **(sizeof(processor) - 1)** automatically calculates the element number of the last element of the array (string). We could have just used the value 3 here. Note that the **sizeof** keyword returns the size of the string including the null-terminating character (4) this is why we need to subtract 1.

12.3.1 Passing an Array to a Function

Because the name of an array is the address of an array, we can only pass the address of an array to a function. It is actually more desirable to pass the address of an array to a function than to copy the entire array to a local array of the function. When a function is passed the address of an array, we are actually passing a pointer to the array to the function. The function can be written to accept the pointer to the array using either pointer notation or array notation.

These two functions can both take an array of type **`char`** as arguments:

```
void Func1(char str[]);
void Func1(char *str);
```

The **`pass_array`** program uses two functions to demonstrate how to pass an array to a function as shown above.

```
#include <stdio.h>
#include "dbgu.h"

void PrintCharArray(char str[]);
void PrintIntArray(int *ptr, int length);

int main(void)
{
    char text[] = "Start pump\r\n";
    int samples[] = {115, 2000, 95, 9 };

    DBGUInit(DBGU_115200);
    PrintCharArray(text);
    PrintIntArray(samples, sizeof(samples) / 4);
    while (1);
}

void PrintCharArray(char str[])
{
    int i = 0;

    while (str[i] != 0) {
        DBGUTx(str[i]);
        i++;
    }
    while (*str != 0) {
        DBGUTx(*str);
        str++;
    }
}

void PrintIntArray(int *ptr, int length)
{
    char str[50];
    int i;

    for (i = 0; i < length; i++) {
```

```
        sprintf(str, "%d: %4d\r\n", i, ptr[i]);
        DBGUTxMsg(str);
    }
    for (i = 0; i < length; i++) {
        sprintf(str, "%d: %4d\r\n", i, *ptr);
        DBGUTxMsg(str);
        ptr++;
    }
}
```

This program defines and initialises two arrays – a character array that is initialised with a string and an array of integers. The **`PrintCharArray()`** function is called and passed the address of the character array **`text`**. The **`PrintIntArray()`** function is called and is passed the address of the integer array **`samples`**. The length of the integer array is also passed to this function.

Each of the functions that are called demonstrates a different way to access an array, but they are essentially doing the same thing. The **`PrintCharArray()`** function accesses the address of the array with a local character pointer that is defined with array notation – **`char str[]`**. The first **`while`** loop in this function uses array notation to print the string out a character at a time. The second **`while`** loop does exactly the same thing, but this time using pointer notation. The loop will only exit if the character that the pointer is pointing to is equal to 0, if the character is not equal to 0 the loop will not terminate – **`(*str != 0)`**. In other words we are checking for the null-terminator in the string and will exit the loop when it is found. To move from character to character using pointer notation, the pointer itself is incremented (`str++`). This increments the address that the pointer is holding i.e. pointing to. If the array was at address 0x20F010, the pointer **`str`** would contain this address after the **`PrintCharArray()`** function was called. After each increment of this address in **`str`**, this pointer would now be pointing to the next element in the array at the next address. This function demonstrates that pointer notation and array notation can be used interchangeably. You can use either one or mix them to achieve the same results.

The **`PrintIntArray()`** function is passed the address of the array which it stores in the local integer pointer **`ptr`**. In the first **`for`** loop in this function, array notation is used with this pointer to access each element in the array that the pointer is pointing to. The second **`for`** loop uses pointer notation to do exactly the same thing as with the **`PrintCharArray()`** function. Again the pointer itself is incremented, which results in it pointing to the next address and therefore the next element in the array. It was necessary to pass the length of this array to this function as there is no other way to determine the number of integers in the array, unlike strings in which the null-terminator can be checked for.

12.4 Strings as Pointers

The **string_ptr** program shows that a string can be initialised as a pointer again showing the similarities between pointer and array notation.

```
#include "dbgu.h"

int main(void)
{
    char *status1 = "Pump started.\r\n";
    char status2[] = "Mixer started.\r\n";

    DBGUInit(DBGU_115200);
    DBGUTxMsg(status1);
    DBGUTxMsg(status2);
    while (1);
}
```

The first string in this program is initialised using pointer notation. In the program it can now be accessed using pointer or array notation. The second string is initialised as we have done before with array notation. It can also be accessed in the program with array or pointer notation. Note that initialising a string as a pointer can only be used with strings. If you wanted to have an array of integers or floating point numbers, you would have to initialise the array with array notation, although you are free in the program to access the array with either notation.

12.5 A Look at the DBGUTxMsg() Function

You now have enough knowledge to fully understand the **DBGUTxMsg()** function that you have been using to send strings out of the embedded system's serial port.

The function is shown below as it is found in **dbgu.c** in the **Drivers** subdirectory of the template files:

```
void DBGUTxMsg(char* msg)
{
    int ch_num = 0;

    /* send characters until null received */
    while (msg[ch_num] != 0) {
        while (!(DBGU_SR & 0x00000002));
        DBGU_THR = (unsigned int)msg[ch_num];
        ch_num++;
    }
}
```

The address of a string is passed to this function and is available in the function in the pointer **msg**. The **while** loop sends one character from the array that **msg** is pointing to out of the serial port at a time. Array notation is used in the loop expression to check for the null-terminating character. Inside the loop, another **while** loop with no body is used to wait until the transmitter is ready before sending a character from the array to the Transmit Holding Register (THR) of the microcontroller's DBGU serial port. The THR register is defined as being 32 bits wide and therefore the character from the array is cast to an **unsigned int**. This puts the character in the lowest byte (LSB) of the THR register which is required in order to send the byte (see the datasheet).

12.6 Multidimensional Arrays

A two dimensional array can be defined in C as follows:

```
int array[3][4];
```

This line tells the compiler to make space in memory for three arrays that are each four integers long. Both the array and each element in the array can be accessed using array notation. To assign a value to the second element of the first array, we would write:

```
array[0][1] = 10;
```

A two dimensional array can be initialised when it is defined as follows:

```
int array[3][4] = {{1, 2, 3, 4}, {11, 12, 13, 14}, {21, 22, 23, 24}};
```

Figure 12.4 shows what this array would look like in memory.

array[0][0]	1	array[1][0]	11	array[2][0]	21
array[0][1]	2	array[1][1]	12	array[2][1]	22
array[0][2]	3	array[1][2]	13	array[2][2]	23
array[0][3]	4	array[1][3]	14	array[2][3]	24

Figure 12.4: A Two Dimensional Array In Memory

The **multi_seq** program uses a two dimensional array to store three sets of patterns that are written to the LEDs. A pattern is displayed on the LEDs continually and can be changed to one of the other patterns by pressing one of the push buttons on the system.

```
#include "at91sam7s.h"

void GetSwitch(void);
void Delay(void);
int prog_num = 0;

int main(void)
{
    unsigned int seq[3][4] =
        {{0x00000009, 0x00000006, 0x0000000F, 0x00000006},
         {0x00000008, 0x00000004, 0x00000002, 0x00000001},
```

```
        {0x00000008, 0x00000009, 0x0000000D, 0x0000000F}};

    int ptrn;
    /* enable the clock of PIOA (Parallel I/O controller A) */
    PMC_PCER = 0x00000004;
    PIO_PER = 0x0008C00F;         /* enable uC pins as PIO */
    PIO_OER = 0x0000000F;         /* enable uC pins as output */
    PIO_ODR = 0x0008C000;         /* enable uC pins as inputs */
    PIO_SODR = 0x0000000F;        /* switch all LEDs off */

    while (1) {
        for (ptrn = 0; ptrn < 4; ptrn++) {
            PIO_CODR = seq[prog_num][ptrn];
            Delay();
            PIO_SODR = seq[prog_num][ptrn];
        }
    }
}

void GetSwitch(void)
{
    if (!(PIO_PDSR & 0x00004000))
        prog_num = 0;
    else if (!(PIO_PDSR & 0x00008000))
        prog_num = 1;
    else if (!(PIO_PDSR & 0x00080000))
        prog_num = 2;
}

void Delay(void)
{
    int del = 500000;

    while (del--) {
        GetSwitch();
    }
}
```

This program is very similar to the **LED_seq** program. It is actually simpler than it looks. The first part of the program inside **main()** defines and initialises a two dimensional array that contains the three sets of patterns that are to be written to the LEDs.

The next few lines before the **while(1)** loop set up port pins PA0 to PA3 as outputs to drive the four LEDs and port pins PA14, PA15 and PA19 as inputs to read the push button switches. You have already set up the port pins as inputs and outputs before, so modify these lines to suit your embedded system if it uses different port pins for its switches and LEDs.

The **while(1)** loop contains a **for** loop that continually writes a pattern to the LEDs from the two dimensional array. You have seen this before with a one dimensional array, except that this time the set of patterns that is output from the two dimensional array is selected by pressing one of the push buttons.

The state (on or off) of the push buttons is checked in the **GetSwitch()** function. This function writes a value to the global variable **prog_num**, with the value depending on which push button is pressed. The **GetSwitch()** function is called from the **Delay()** function as this is where most of the microcontroller's processing time is used up and checking the switches here will make the system more responsive i.e. you don't have to wait with the switched pressed in until the delay routine is finished in order for the system to read the switch.

The pattern selected by each switch press is done in the array as shown in **figure 12.5**. The global variable **prog_num** can be seen by every function in the source file so does not have to be passed to the function as an argument or returned by the function.

Pushing a switch changes the index in the array and selects a "sub-array" 0, 1 or 2.

```
PIO_CODR = seq[prog_num][ptrn];
```

Each element in the "sub-array" is accessed by the for loop counting from 0 to 3.

Figure 12.5: Accessing Elements in a Two-Dimensional Array

The next example, **status_array**, shows how an array of strings is initialised and used for displaying status messages of an imaginary embedded controller system.

```
#include "dbgu.h"

#define NUM_MSGS    5

int main(void)
{
    int msg_num = 0;
    char status[NUM_MSGS][25] = {"Belt Jam Alarm.\r\n",
                                 "Conveyor Running.\r\n",
                                 "Conveyor Stopped.\r\n",
                                 "Out of Paper.\r\n",
                                 "System Ready.\r\n"};
    DBGUInit(DBGU_115200);
    for (msg_num = 0; msg_num < NUM_MSGS; msg_num++) {
        DBGUTxMsg(status[msg_num]);
    }
    while (1);
}
```

The two dimensional array **status** is initialised with strings which are of course themselves arrays. The size of each string can be 25 characters long including the null-terminating character. For strings that are smaller than this, the extra bytes are wasted.

The program uses a **for** loop to send each string out of the serial port to be displayed on HyperTerminal. The number that the **for** loop is counting through is used as an index into the two dimensional array, resulting in the address of each string in the array being passed to the **DBGUTxMsg()** function. Because the **DBGUTxMsg()** function is sent the address of a string, it will send each character in the string out of the serial port until it finds the null-terminating character. It does not care whether the string is part of a multidimensional array or where it exists in memory, it just needs the address of a string.

A more efficient method can be used to initialise an array of strings. You have seen that memory is wasted in the previous example by having to specify a fixed length for each string, even though every string does not make use of the entire length. Initialising an array of pointers to strings will not waste any bytes. A modified version of the previous program, called **status_point** uses an array of pointers to strings. The program listing is not shown as the only thing that changes in the program is the line:

```
char status[NUM_MSGS][25]
```

Changes to:

```
char *status[NUM_MSGS]
```

The changed line simply states that we want an array (**status[NUM_MSGS]**) and that data stored in the array is going to be of the type pointers to characters (**char ***). Now the line:

```
DBGUTxMsg(status[msg_num]);
```

passes each pointer in the array that is pointing to the first character of each string to the **DBGUTxMsg()** function. The program does exactly the same thing as the previous one, only it saves some memory that was previously wasted.

12.7 Exercises

1. Write a program that sends a message to HyperTerminal that says which button on the embedded system was pressed. The messages must be stored as an array of pointers to strings. Use an index to access each message.

12.7.1 Solutions

1. **sw_message**:

```
#include "at91sam7s.h"
#include "dbgu.h"

int GetSwitch(void);

int main(void)
{
    char *sw_msgs[] = {"SWITCH 1\r\n",
                       "SWITCH 2\r\n", "SWITCH 3\r\n"};
    int index;

    DBGUInit(DBGU_115200);
    /* enable the clock of PIOA (Parallel I/O controller A) */
    PMC_PCER = 0x00000004;
    PIO_PER = 0x0008C000;       /* enable uC pins as PIO */
    PIO_ODR = 0x0008C000;       /* enable uC pins as inputs */

    while (1) {
        index = GetSwitch();
        if (index >= 0) {
            DBGUTxMsg(sw_msgs[index]);
        }
    }
}

int GetSwitch(void)
{
    int sw_num = -1;

    if (!(PIO_PDSR & 0x00004000))
        sw_num = 0;
    else if (!(PIO_PDSR & 0x00008000))
        sw_num = 1;
    else if (!(PIO_PDSR & 0x00080000))
        sw_num = 2;

    return sw_num;
}
```

In this solution, **GetSwitch()** will return -1 when called if no switch is pressed or else 0 to 2 if a switch is pressed. The **if** statement in the **while(1)** loop checks to see if a switch was pressed by checking for a number that is bigger than -1 i.e. 0 or greater.

12.8 Summary

- Arrays are sets of sequential data types.
- Each element in an array can be accessed using array notation.

- Strings are arrays of characters that are null terminated.
- The address of an array in C is the name of the array.
- Arrays can be accessed using array or pointer notation.

13 Bit Manipulation and Logical Operators

What you will do in this chapter:

- See how to change individual bits inside a variable by using bitwise operators
- Move the contents of a variable left or right using the shift operators
- Use logical operators to combine expressions

13.1 Bit Manipulation with Bitwise Operators

Bitwise operators enable a programmer to modify individual bits in a memory location and can be used to operate on bytes or larger groups of bits. The C programming language has four Boolean bitwise operators and two shift operators as shown in **table 13.1**. The bitwise **AND** operator (`&`) should be familiar to you.

Boolean Bitwise Operators			
Operator	**Operation**	**Example**	**Result**
`&`	AND	`1101 & 0111`	`0101`
`|`	OR	`1101 | 1001`	`1101`
`^`	XOR (exclusive OR)	`1101 ^ 1001`	`0100`
`~`	Not (Compliment)	`~1101`	`0010`

Shift Operators			
Operator	**Operation**	**Example**	**Result**
<<	Left Shift	`0010 << 1`	`0100`
>>	Right Shift	`0010 >> 1`	`0001`

Table 13.1: C Boolean Bitwise and Shift Operators

Windows calculator will again come in handy when learning about bitwise operators as it has all of the bitwise operators available except the right shift operator. **Figure 13.1** shows the Windows calculator and the equivalent C bitwise operators.

The bitwise operators can be tested with the calculator in binary or hexadecimal mode and it is possible to change between these modes at any time while performing an operation. The keystrokes used to calculate the examples in **table 13.1** on the calculator are shown below from top to bottom (first put the calculator into binary mode):

1101 And 111 =

Note that the 0 place holder to the left of the second number is left out when entering the second number and is not displayed on the calculator in the result i.e. the result displayed is **101** and not **0101**.

1101 Or 1001 =

1101 Xor 1001 =

1101 Not

The result of this operation on the calculator turns all of the invisible 0 place holders to the left of the number into 1s and that is why the number is filled with 1s to its left.

10 Lsh 1 =

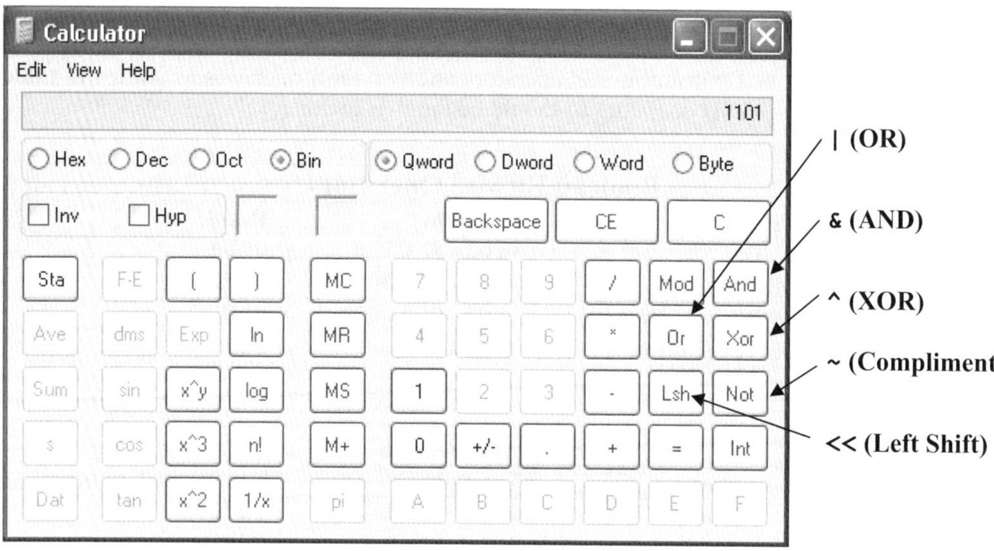

Figure 13.1: Bitwise Operations on the Windows Calculator

In order to understand what the Boolean bitwise operators do, we need to look at the truth tables of these operators. The truth tables are shown in **figure 13.2**. The truth tables show the output of each bitwise operator when it is performed on the input bit(s).

& (AND)		
Inputs		**Output**
0	0	0
0	1	0
1	0	0
1	1	1

| (OR)		
Inputs		**Output**
0	0	0
0	1	1
1	0	1
1	1	1

^ (XOR)		
Inputs		**Output**
0	0	0
0	1	1
1	0	1
1	1	0

~ (NOT)	
Input	**Output**
0	1
1	0

Figure 13.2: Truth Tables for the Boolean Bitwise Operators

The **NOT** operator simply inverts each bit in a byte i.e. it changes all of the 0s to 1s and 1s to 0s.

In C, the other three operators are performed on two or more memory locations of length 8 bits (**`char`**), 16 bits (**`short`**) or 32 bits (**`int`** or **`long`**). Let's take the **OR** operator as an example and use it to operate on two bytes. In C the **OR** operator can be used as follows:

```
unsigned char byte1 = 0xC1;
unsigned char byte2 = 0x13;
unsigned char result;

result = byte1 | byte2;
```

The above example **ORs** two variables together and saves the result in a variable called **`result`**. The example could also use the two numbers directly i.e. it could be used to **OR** two constants together and store the result in a variable. **Figure 13.3** shows how the **OR** operation is done.

	Bit 7	Bit 6	Bit 5	Bit 4	Bit 3	Bit 2	Bit 1	Bit 0
byte1	1	1	0	0	0	0	0	1
byte2	0	0	0	1	0	0	1	1
result	1	1	0	1	0	0	1	1

Bit 0: 1 and 1 are <u>inputs</u> in the truth table, 1 is the <u>output</u> in the truth table
Bit 1: 0 and 1 are <u>inputs</u> in the truth table, 1 is the <u>output</u> in the truth table
Bit 2: 0 and 0 are <u>inputs</u> in the truth table, 0 is the <u>output</u> in the truth table
etc.... for each bit

Figure 13.3: The Bitwise OR Operation on Two Bytes

Each bit of **`byte1`** is **ORed** with the corresponding bit in **`byte2`** from top to bottom as shown in **figure 13.3**. Using the **OR** truth table in **figure 13.2**, bit 0 of each byte are **ORed** together i.e. **1 OR 1 = 1**. The result is stored in the corresponding bit of **`result`** i.e. bit 0 of **`result`**. Bit 1 is done in the same way: **0 OR 1** = **1**. Bit 2: **0 OR 0** = **0**. And so on for each bit from 0 to 7. The same operation would be done for the **AND** and **XOR** if they were used instead of the **OR** operator, the only difference would be the truth table used and the result produced.

13.1.1 Why Do We Need Bitwise Operators?

Bitwise operators make hardware programming easy to do. There are many cases where a single bit needs to be changed in order to operate hardware such as an output. You have already seen the **AND** operator used to test a single bit in a word that was read back from a hardware register. Before looking at the left and right shift operators, we will look at some example programs that use the Boolean bitwise operators.

When switching LEDs on and off using the PIO controller of the AT91SAM7S microcontroller, we have been using the PIO_CODR and PIO_SODR for clearing and setting bits on the PIO port. There is another register, the PIO_ODSR (PIO Output Data Status Register), that allows direct access to the PIO pins of the microcontroller by single write operations to this register. When using the clear and set registers, we had to clear away previous data written to switch all the LEDs off before switching the next pattern of LEDs on. A single write to the PIO_ODSR will set the state of the pins to the logic levels in this register. Writing a 0 to the set or clear registers had no effect on the output. Writing a 0 to the PIO_ODSR register will switch the corresponding pin low and writing a 1 to this register will switch the corresponding pin high.

Before we can write a value to the PIO_ODSR that can be displayed on our LEDs, we need to write to the PIO_OWER (PIO Output Write Enable Register). Writing 1s to the various bits in this register will enable the corresponding pins on the PIO port to be written to by the PIO_ODSR, the other bits will be "masked out" (the corresponding bits that 1 is not written to).

The AT91SAM7S-EK evaluation board has LEDs attached to port pins PA0 to PA3, so a value of 0x0000000F will need to be written to the PIO_OWER to enable these pins to be written to by the PIO_ODSR. The **`port_write`** program will show how this is done.

```
#include "at91sam7s.h"
void Delay(void);

int main(void)
{
    /* configure PIO lines for LEDs on PA0 to PA3 */
    PIO_PER = 0x0000000F;         /* enable uC pins as PIO */
    PIO_OER = 0x0000000F;         /* enable uC pins as output */
    PIO_OWER = 0x0000000F;        /* allow direct writing to port bits */

    while (1) {
        PIO_ODSR = 0x00000005;
        Delay();
        PIO_ODSR = 0x0000000A;
        Delay();
```

```
    }
}
```

This program enables the pins PA0 to PA3 as outputs as we have done before. You will need to change the values written in the program to enable the LED pins on your board should they differ. The PIO_OWER is also written to in order to allow direct access by the PIO_ODSR. Note that only the lower four bits of this register that correspond to port pins PA0 to PA3 are set. Only values written to bits 0 to 3 will be output on the corresponding port pins when values are written to PIO_ODSR because of this. If any other bit is written to in the PIO_ODSR, it will be masked out and not affect the corresponding pin.

In the **`while(1)`** loop, two values are written to port pins PA0 to PA3 with a delay in-between. These values will be displayed on the LEDs when written. Previously when using set and clear registers we would have written the first value, called the delay function and then had to clear the value that we wrote, then write the new value, call the delay function again and then clear the second value. With the PIO_ODSR, we do this in a single step per value written. Note that the delay function is not included in the listing as it is the same one that has previously been used.

This example did not use any bitwise operations, but the next example will set up the LED pins on the PIO controller in the same way and then demonstrate the use of the bitwise operators. This example is called **`bitwise_port`**.

```
#include "at91sam7s.h"
void Delay(void);

int main(void)
{
    /* configure PIO lines for LEDs on PA0 to PA3 */
    PIO_PER = 0x0000000F;        /* enable uC pins as PIO */
    PIO_OER = 0x0000000F;        /* enable uC pins as output */
    PIO_OWER = 0x0000000F;       /* allow direct writing to port bits */

    PIO_ODSR = 0x00000005;
    while (1) {
        PIO_ODSR = PIO_ODSR & 0xFFFFFFFE;
        Delay();
        PIO_ODSR = PIO_ODSR | 0x00000001;
        Delay();
    }
}
```

This program is similar to the previous program, but uses the bitwise **AND** and **OR** operators to switch an LED on and off. The object of the program is to demonstrate how to toggle (switch on and off) a single bit in a register without affecting the other bits in the register. Because the register changes the states of the port pins, the program will toggle a single LED without affecting the state (on or off) of the other LEDs on the port.

An initial value of 0x5 is written to the four LEDs before the **`while(1)`** loop is entered. This value switches two LEDs on and leaves two LEDs off. We now want to change the state of a single LED on the port without affecting the state of the other LEDs on the port. This is done using bitwise operators inside the **`while(1)`** loop. The bitwise operators are used to clear and set the LSB of the PIO_ODSR and thereby switch on and off the LED on port pin PA0. The bit is cleared by **ANDing** it with 0 while **ANDing** everything else in the register with 1. This results in the LSB changing to 0 and all the other bits in the register staying the same. We have thus manipulated a single bit in this 32 bit register.

After calling the delay routine, the bitwise **OR** operator is used to set the LSB of the PIO_ODSR. This is done by **ORing** the PIO_ODSR with 1, resulting in the LSB of this register changing to 1 and all of the other bits in the register remaining the same (they are all **ORed** with 0 and therefore stay the same). **Figure 13.4** shows what is happening in the register. Only the lowest 8 bits of the register are shown.

Bit Number	**31**	**30**	**29**	**....**	**7**	**6**	**5**	**4**	**3**	**2**	**1**	**0**
PIO_ODSR	0	0	0	...	0	0	0	0	0	1	0	1
Constant	1	1	1	...	1	1	1	1	1	1	1	0
Result	0	0	0	...	0	0	0	0	0	1	0	0

Bitwise AND (&) Operation

Bit Number	**31**	**30**	**29**	**....**	**7**	**6**	**5**	**4**	**3**	**2**	**1**	**0**
PIO_ODSR	0	0	0	...	0	0	0	0	0	1	0	0
Constant	0	0	0	...	0	0	0	0	0	0	0	1
Result	0	0	0	...	0	0	0	0	0	1	0	1

Bitwise OR (|) Operation

Note: The Result is stored in PIO_ODSR after each operation.

Figure 13.4: The AND and OR Operations in the `bitwise_port` Program

In this program a read-modify-write is performed on the PIO_ODSR register when the bitwise operators are used. For example in the line:

```
PIO_ODSR = PIO_ODSR & 0xFFFFFFFE;
```

The contents of the PIO_ODSR are first read by the microcontroller – this moves the value in the PIO_ODSR into an internal register of the microcontroller (read). The **AND** operation is then performed (modify) and the result is written back to the PIO_ODSR (write). There is a shorthand way of writing the above line:

```
PIO_ODSR &= 0xFFFFFFFE;
```

The **`&=`** is known as a bitwise assignment operator. This operator **ANDs** the two values to the left and right of it together and then stores the result in the variable on the left of it. **Table 13.2** at the end of this section shows the C assignment operators.

The next program shows the use of the bitwise **NOT** operator, also known as the **compliment** operator. On the AT91SAM7S-EK evaluation board, the LEDs are wired to be switched on by writing a 0 to the port and off by switching a 1 to the port. This hardware configuration is known as current sinking. With this configuration, the value that you write to the port gives the opposite of what you would expect when it is displayed on the LEDs. Writing **1100** to the port will display as if you had written **0011**, as the last two LEDs on the port are lit up instead of the two to the left of them. The **not_port** program uses the bitwise **NOT** operator to invert (change 0s to 1s and 1s to 0s) the value written to the port and display the expected values on an evaluation board that has LEDs that are wired as current sinking.

```
#include "at91sam7s.h"
void Delay(void);

int main(void)
{
    int count = 0;
    /* configure PIO lines for LEDs on PA0 to PA3 */
    PIO_PER = 0x0000000F;         /* enable uC pins as PIO */
    PIO_OER = 0x0000000F;         /* enable uC pins as output */
    PIO_OWER = 0x0000000F;        /* allow direct writing to port bits */

    while (1) {
        PIO_ODSR = ~count;
        Delay();
        count++;
        if (count > 0x0F)
            count = 0;
    }
}
```

This program writes a count to the LED port. Note that if the LEDs on your embedded system are not attached to pins PA0 to PA3, the program will not work properly or at all. You will need to wait until you can use the left shift operator before you will be able to modify this program to run.

When the count is written to the LEDs, it is inverted with the **NOT** operator in the line:
`PIO_ODSR = ~count;`

The result of this operation is a correctly displayed series of binary count values if the LEDs are wired to the port pins as current sinking.

13.1.2 The Left and Right Shift Operators

The shift operators will shift the entire contents of a value by the specified number of bits. The two shift operators can move the bits either to the left (**<<**) or to the right (**>>**). The next program will use the shift operators to create a Knight Rider display on the LEDs. Here is the **knight_disp** program:

```
#include "at91sam7s.h"

void Delay(void);
```

```
int main(void)
{
    int move_left = 1;
    unsigned char shift_val = 1;
    /* configure PIO lines for LEDs on PA0 to PA3 */
    PIO_PER = 0x0000000F;        /* enable uC pins as PIO */
    PIO_OER = 0x0000000F;        /* enable uC pins as output */
    PIO_OWER = 0x0000000F;       /* allow direct writing to port bits */
    while (1) {
        if (move_left) {
            shift_val = shift_val << 1;
            if (shift_val & 0x08)
                move_left = 0;
        }
        else {
            shift_val = shift_val >> 1;
            if (shift_val & 0x01)
                move_left = 1;
        }
        PIO_ODSR = ~shift_val;
        Delay();
    }
}
```

This program uses the **move_left** variable as a "flag" that is used to flag when the value displayed on the LEDs is to be left shifted and when it is to be right shifted. By shifting a single bit that is set, and then writing out this value to the LEDs, we achieve the effect of an LED appearing to run to the left and then to the right – the "Knight Rider" display.

The value to write to the LEDs is stored in the variable **shift_val** and is initialised to the value of 1 i.e. this is a single byte and is initialised to the binary value of 00000001B. The flag variable, **move_left** is also initialised to the value of 1 (true). When a variable is used as a flag, it is checked in a program for being either true (contains a non-zero value) or false (contains 0).

Upon entry into the **while(1)** loop, the **move_left** flag causes the code in the body of the **if** statement to be executed as this flag evaluates to true. The value in **shift_val** is shifted to the left by 1 bit and the result is stored back in **shift_val**. The value will therefore have changed from 00000001B to 00000010B. A check is done after this to see if the value 0x08 has been reached i.e. 00001000B. This check is done to see if the fourth bit from the right of this variable is set because we are writing to only four LEDs and if it was not done, the bit would be shifted too far to the left and would not display on the LEDs.

The **shift_val** variable is written to the PIO_ODSR so that it will be displayed on the LEDs. Note that this value is inverted when written to the register to compensate for the current sinking configuration of the LEDs on the evaluation board. If the value was not inverted with the **NOT** logical operator, the LEDs would all be on and the "moving" LED would be off.

After the delay function is called, the **`if`** statement is evaluated again. The **`shift_val`** variable will be shifted left again in the body of the **`if`** statement as before until the value becomes 0x08 (**`00001000B`**) the **`move_left`** flag will then be cleared and the body of the **`else`** statement will be run on the next pass through the loop.

The code in the body of the **`else`** statement does exactly the same as in the **`if`** statement, except that the value in **`shift_val`** is now shifted right by one bit and a check is done for the value 0x01 (**`00000001B`**) before changing the **`move_left`** flag to true (1).

When **`shift_val`** is written to the PIO_ODSR, we are actually writing an 8 bit value to a 32-bit memory location. This is not a problem as this byte will be written to the least significant byte of the 32-bit memory location i.e. bits 0 to 7 of the byte will be written to bits 0 to 7 of the 32-bit location. This is exactly what we want as the lower 4 bits of **`shift_val`** contain the value that we want to write to the LEDs and the LEDs are "wired" to the lowest 4 bits of the PIO_ODSR.

13.1.3 The C Assignment Operators

There is a shorthand method for performing arithmetic and bitwise logical operations on data in C and storing the result in one of the variables that is being operated on. We saw this with the bitwise **AND** operator earlier in this chapter. As an example, in the previous program example, the following line:

```
shift_val = shift_val << 1;
```

could be changed to:

```
shift_val <<= 1;
```

Both of these lines do exactly the same thing, the second one is just a shorter way of writing it. The value inside **`shift_val`** is moved by 1 bit to the left and the result is stored back in **`shift_val`**. **Table 13.2** shows all of the assignment operators in C.

<table>
<tr><th>Operator</th><th>Alternate Notation</th><th>Meaning</th></tr>
<tr><td>=</td><td>A = B</td><td>The value in B is assigned to variable A</td></tr>
<tr><td>+=</td><td>A = A + B</td><td>A is assigned the sum of A and B</td></tr>
<tr><td>-=</td><td>A = A – B</td><td>A is assigned the difference between A and B</td></tr>
<tr><td>*=</td><td>A = A * B</td><td>A is assigned the value of A multiplied by B</td></tr>
<tr><td>/=</td><td>A = A / B</td><td>A is assigned the value of A divided by B</td></tr>
<tr><td>%=</td><td>A = A % B</td><td>A is assigned the value of the modulus of A and B</td></tr>
<tr><td>&=</td><td>A = A & B</td><td>A is assigned the value of A ANDed with B</td></tr>
<tr><td>^=</td><td>A = A ^ B</td><td>A is assigned the value of A XORed with B</td></tr>
<tr><td>|=</td><td>A = A | B</td><td>A is assigned the value of A ORed with B</td></tr>
<tr><td><<=</td><td>A = A << x</td><td>A is assigned the result of left shifting A by x bits</td></tr>
<tr><td>>>=</td><td>A = A >> x</td><td>A is assigned the result of right shifting A by x bits</td></tr>
</table>

Table 13.2: The Assignment Operators in C

13.2 Logical Operators

The logical operators in C are represented by symbols similar to the bitwise operators, but instead of operating on single bits are used to operate on the results of expressions used in decisions and loop statements. We will see what this means when we look at some examples. **Table 13.3** shows the logical operators in C. From the table you should recognise the **logical NOT** operator (`!`) as we have used it before in example programs.

<table>
<tr><th>Operator</th><th>Description</th><th>Example</th><th>Result</th></tr>
<tr><td>&&</td><td>Logical AND</td><td><code>A && B</code></td><td>True if A AND B are both true, else false</td></tr>
<tr><td>||</td><td>Logical OR</td><td><code>A || B</code></td><td>False if both A and B are false, else true</td></tr>
<tr><td>!</td><td>Logical NOT</td><td><code>!A</code></td><td>True if A is false, False if A is true</td></tr>
</table>

Table 13.3: The C Language Logical Operators

The logical operators use the same truth tables as the bitwise operators in **figure 13.2**. The next example program, **`or_keys`**, shows how the logical operators are used.

```
#include "dbgu.h"

int main(void)
{
    char rx_data;

    DBGUInit(DBGU_115200);
    while (1) {
        if (DBGUCheckForRxData()) {
            rx_data = DBGURx();
            if (rx_data == 'a' || rx_data == 'A') {
                DBGUTxMsg("\r\nSystem Activated...\r\n");
            }
            else if (rx_data == 'd' || rx_data == 'D') {
```

```
                DBGUTxMsg("\r\nSystem Deactivated...\r\n");
            }
        }
    }
}
```

This program will send a message out of the serial port to be displayed on HyperTerminal if the 'a' or 'A' character is sent to it and it will send a different message if the 'd' or 'D' character is sent. It does not matter whether the Caps Lock key on the keyboard is active or not as the program will perform the same action if the correct upper or lower case character is pressed. This is achieved by using the logical **OR** operator (||).

Let's examine the **`if`** statement to see how this works:

`if (rx_data == 'a' || rx_data == 'A')`

If a character other than 'a' or 'A' has been received and stored in the **`rx_data`** variable, the if statement will evaluate each of the comparative statements which will both result in a value of 0 (false):

`if (0 || 0)`

From the truth table 0 **OR** 0 = 0, so the **`if`** statement will evaluate to false and the code in the body of this statement will not be run.

Should the data received and stored in **`rx_data`** be the character 'a', the **`if`** statement will evaluate as follows:

`if (1 || 0)`

This is because the first comparative statement evaluates to true (1) as **`rx_data`** is equal to 'a' and the second one evaluates to false. From the **OR** truth table, 1 **OR** 0 = 1. The **`if`** statement will therefore evaluate to true and the code in the body of this statement will be run. The same happens if **`rx_data`** contains the character 'A', from the truth table: 0 **OR** 1 = 1, so the **`if`** statement will also evaluate to true in this case and the code in the body of this statement will be run.

In this program it is not possible for the combination 1 **OR** 1 = 1 from the truth table to occur as the variable **`rx_data`** cannot contain both the characters 'a' and 'A' at the same time.

The **`and_button`** program uses a mix of bitwise operators and logical operators to test if more that one switch is pressed on the embedded system.

```
#include "at91sam7s.h"

int main(void)
{
    unsigned int sw_val;
    /* enable the clock of the PIO controller */
    PMC_PCER = 0x00000004;
```

```
    /* configure PIO lines for LEDs on PA0 to PA3 */
    PIO_PER = 0x0000C00F;        /* enable uC pins as PIO */
    PIO_OER = 0x0000000F;        /* enable uC pins as output */
    PIO_ODR = 0x0000C000;        /* pins used as input */
    PIO_OWER = 0x0000000F;       /* allow direct writing to port bits */
    while (1) {
        sw_val = PIO_PDSR;
        sw_val = ~sw_val;
        if (sw_val & 0x00008000 && sw_val & 0x00004000) {
            PIO_ODSR = ~0x0000000F;
        }
        else {
            PIO_ODSR = 0x0000000F;
        }
    }
}
```

This program enables the microcontroller to drive four LEDs on port pins PA0 to PA3 and enables the buttons on PA14 and PA15 to be read. In the `while(1)` loop, the states of the buttons are read and only if both buttons are pressed, will all four LEDs be switched on. If either of the buttons or both are released, the four LEDs will be switched off.

In the `while(1)` loop, the microcontroller port register is read which will contain the state of the two switches or buttons that are being monitored. The value in this register is stored in the local variable `sw_val`. Because the switches are wired on the embedded system so that they are read as 1's when open and 0's when pressed, the value in `sw_val` is inverted using the bitwise **NOT** operator (`~`). The corresponding bits in this variable will now be 1 if a switch is pressed and 0 if it is released.

The `if` statement evaluates the state of the switches that are now captured in the `sw_val` variable. The bitwise **AND** (`&`) operations are first evaluated and then these results are evaluated together using the logical **AND** operator (`&&`).

In the `if` statement:

```
if (sw_val & 0x00008000 && sw_val & 0x00004000)
```

`sw_val & 0x00008000` is evaluated first and will evaluate to either true or false depending if the switch on PA15 is pressed or not. Next, `sw_val & 0x00004000` is evaluated to see if it is true or false depending on if the switch on PA14 is pressed or not. These results are then evaluated using the **AND** truth table with the result from this truth table allowing the code in the body of the `if` statement to be run if it evaluates to true (1):
0 AND 0 = 0, neither switches are pressed, so the `if` statement evaluates to false.
0 AND 1 = 0, the switch on PA14 is pressed, the `if` statement evaluates to false.
1 AND 0 = 0, the switch on PA15 is pressed, the `if` statement evaluates to false.
1 AND 1 = 1, both switches are pressed, the `if` statement evaluates to true.

13.3 Operator Precedence

Operator precedence refers to the order in which expressions with multiple operators are evaluated. For example in the following C statement:

```
x = 5 + 3 * 2;
```

Is 3 added to 5 first and then the result multiplied by 2 (x = 16)? Or is 3 multiplied by 2 and the result added to 5 (x = 11)? C follows the same rules as mathematics in this case where the multiplication sign takes precedence over the plus sign so that 3 is first multiplied by 2 and the result is added to 5 so that x is assigned the result of 11.

Table 13.4 contains the operator precedence of the C operators. The only things that you have not seen so far that are in this table are the dot operator (`.`) and the structure pointer (`->`).

<table>
<tr><th>Precedence</th><th>Operator</th></tr>
<tr><td>1</td><td>() [] . -></td></tr>
<tr><td>2</td><td>++ -- ! ~ + - * & sizeof</td></tr>
<tr><td>3</td><td>cast operator</td></tr>
<tr><td>4</td><td>* / %</td></tr>
<tr><td>5</td><td>+ -</td></tr>
<tr><td>6</td><td><< >></td></tr>
<tr><td>7</td><td>< <= > >=</td></tr>
<tr><td>8</td><td>== !=</td></tr>
<tr><td>9</td><td>&</td></tr>
<tr><td>10</td><td>^</td></tr>
<tr><td>11</td><td>|</td></tr>
<tr><td>12</td><td>&&</td></tr>
<tr><td>13</td><td>||</td></tr>
<tr><td>14</td><td>? : (conditional operator)</td></tr>
<tr><td>15</td><td>= += -= *= /= %= &= ^= |= <<= >>=</td></tr>
<tr><td>16</td><td>, (comma operator)</td></tr>
</table>

Table 13.4: Operator Precedence

When more than one operator is used in a C statement, they will be evaluated in the order that they appear in this table from top to bottom. An example from our previous program:

```
if (sw_val & 0x00008000 && sw_val & 0x00004000)
```

The bitwise **AND** operator (`&`) appears before the logical **AND** operator (`&&`) in the table, so the bitwise **ANDs** in the above line will be evaluated before the logical **AND**. Parentheses can be used to change the order of precedence as well as make it clearer what you as the programmer are intending. For example, the above line could be written as:

```
if ((sw_val & 0x00008000) && (sw_val & 0x00004000))
```

This shows that it is intended to first evaluate the bitwise **AND** operators as everything inside parentheses is evaluated first.

With our mathematical example:

```
x = 5 + 3 * 2;
```

We can write this instead:

```
x = 5 + (3 * 2);
```

This shows that we intend to multiply 3 by 2 first and then add 5 to the result before assigning the final result to **x**. We can also use parentheses to change the order of evaluation as follows:

```
x = (5 + 3) * 2;
```

This forces 5 to be added to 3 first, as the parentheses have higher precedence than the multiplication sign. The result of 5 + 3 (8) is then multiplied by 2.

13.4 Exercises

1. Use the bitwise AND operator and the shift operator to check if each bit in a byte is a 0 or a 1. Use the results of the check to create the binary equivalent of the number and send it to be displayed on HyperTerminal.

13.4.1 Solutions

1. num2bin:

```
#include "dbgu.h"

int main(void)
{
    unsigned char num = 0xA3;
    unsigned char mask = 0x80;
    char str[20];
    int str_index;

    DBGUInit(DBGU_115200);

    /* count through each bit in the byte  and build *
     * a string that represents the binary number in *
     * the byte                                      */
    for (str_index = 0; str_index < 8; str_index++) {
        if (mask & num) {              /* check for 1 */
            str[str_index] = '1';    /* bit is a 1 */
        }
        else {
            str[str_index] = '0';    /* bit is a 0 */
        }
```

```
        mask >>= 1;  /* for checking next bit in byte */
    }
    str[str_index] = 0;       /* null terminate the string */
    DBGUTxMsg(str);
    while (1);
}
```

This program uses a **for** loop to count through each of the bits in the byte that we are going to display in binary format. A variable is set up as a mask with its most significant bit set. The program uses the mask to check whether each bit in **num** is a 1 or a 0 by shifting the mask by 1 bit to the right after each check. In other words it is used to check the state of bit 7, bit 6, bit5, etc. and finally bit 0. A string is built up of ASCII characters '1' and '0' after each bit is checked. The string is null terminated when the **for** loop exits and then sent out of the serial port for display on HyperTerminal.

13.5 Summary

- Bitwise operators can be used to change the state of individual bits in memory and check the state of bits.

- The shift operators in C can be used to shift the contents of a variable or memory location left or right by a specified number of bits.

- Logical operators are used to combine more than one expression.

- Operator precedence in C determines what is evaluated first. Operator precedence can be changed by using parentheses.

14 More Hardware Programming

What you will do in this chapter:

- Use the hardware timer counter in the AT91SAM7S to make accurate time delays.
- Read external analogue values using the analogue to digital converter.
- Setup and use interrupts in a program.

We have covered most of the C language up to now. This chapter will deal with using this knowledge to program more hardware devices on the microcontroller such as timers and the analogue to digital converter (ADC). After learning the C language, moving from one embedded system to another or one microcontroller architecture to another is all about learning how the new hardware works and then programming it. We will wrap up in the chapter that follows by covering the remaining parts of the C language that are more advanced and/or less used.

Because this is a book on learning the C language, we will not cover every detail of the microcontroller hardware, but rather demonstrate how various hardware devices are set up and used. A separate book that covers your microcontroller of choice and describes every aspect of the hardware in detail would more appropriately cover these topics. We are not, however isolated from the specific hardware on the microcontroller of choice for this book. Because hardware architectures differ so much between microcontrollers, this chapter has to look at the specific hardware on the AT91SAM7S microcontroller family and we can not talk about setting up timers, etc. in general terms across microcontrollers from different manufacturers.

14.1 The AT91SAM7S Timer Counter

The timer counters (TC) in the AT91SAM7S microcontrollers can be used for multiple purposes such as event counting, pulse generation, delay timing and more. You can read more about the timer counter in chapter 33 of the AT91SAM7S datasheet. The next example program called **`timer`** will use one of the timer counters (TC0) for generating a delay period much like the **`Delay()`** function that we have used in previous programs. Using the timer to generate a delay is a much more accurate method than using a software delay routine.

```
#include <stdio.h>
#include "at91sam7s.h"
#include "dbgu.h"

void TimeDelay(unsigned int ms_delay);

int main(void)
```

```
{
    char str[50];
    unsigned int time = 0;
    /* enable the clock of the TC0 module through the PMC */
    PMC_PCER = 0x00001000;
    /* setup TC0 registers */
    /* waveform mode, clocksource = 5 (MCLK / 1024)*/
    TC0_CMR = 0x0000C004;
    TC0_RC = 0x0000002F;        /* RC Register value for 1ms */
    DBGUInit(DBGU_115200);

    while (1) {
        sprintf(str, "%010d\r", time);
        DBGUTxMsg(str);
        TimeDelay(1000);
        time++;
    }
}

void TimeDelay(unsigned int ms_delay)
{
    TC0_CCR = 0x00000005;       /* start the timer */
    while (ms_delay) {
        if (TC0_SR & 0x00000010)
            ms_delay--;
    }
}
```

Looking at the microcontroller datasheet for information on programming one of the on-chip devices can be rather overwhelming at first. The section in chapter 33 that covers the timer counters has a lot of information, most of which you will not need, depending on which mode you want to set the timer counter up in. There are a lot of choices as the timer counter has several timer modes and can be used for counting as well. It can also be set up to be activated by external signals on the microcontroller's port pins. The example program sets timer counter 0 (TC0) up to time 1ms intervals. A function makes use of the timer and is called in the program to time 1 second intervals. The program sends a count in seconds out of the serial port every second. Let's see how the program works.

Before any of the timer counter units in the microcontroller can be used, a clock signal must be connected to it from another piece of hardware on the microcontroller called the Power Management Controller (PMC). The PMC is explained in chapter 25 of the datasheet. The clock signal is routed by setting a single bit in the PMC Peripheral Clock Enable Register (PMC_PCER). This is done at the beginning of **main()**:

```
PMC_PCER = 0x00001000;
```

Two TC0 registers are used to set up the timer in the mode that we want for timing. You can refer to the datasheet to see which bits are being set by the program. Each register that is written to is explained below. The microcontroller has 3 timers, TC0 to TC2, so all of the register names starting with TC0 refer to timer counter 0 and are defined in the **at91sam7s.h** header file.

```
TC0_CMR = 0x0000C004;
```

This is the Channel Mode Register. A value of **100B** (0x4) is written to the TCCLKS bit field. A bit field refers to a group of bits in a register. If you look in the datasheet or **figure 14.1**, you will see that there are 3 bits grouped together called TCCLKS. There is also a table in the datasheet that shows that writing **100B** to this bit field will select TIMER_CLOCK5. TIMER_CLOCK5 is a clock signal that has a timing period equal to the master clock MCK divided by 1024. This is chosen as the master clock itself is running at a very high speed and we want to run the timer in milliseconds, so need a lower clock speed. Dividing the master clock by 1024 provides the timer with a slower clock speed.

31	30	29	28	27	26	25	24
0	0	0	0	0	0	0	0
–	–	–	–	–	–	–	–
23	**22**	**21**	**20**	**19**	**18**	**17**	**16**
0	0	0	0	0	0	0	0
–	–	–	–	LDRB		LDRA	
15	**14**	**13**	**12**	**11**	**10**	**9**	**8**
1	1	0	0	0	0	0	0
WAVE	CPCTRG	–	–	–	ABETRG	ETRGEDG	
7	**6**	**5**	**4**	**3**	**2**	**1**	**0**
0	0	0	0	0	1	0	0
LDBDIS	LDBSTOP	BURST		CLKI	TCCLKS		

Figure 14.1: The Timer Counter Channel Mode Register (TC_CMR)

The CPCTRG bit is set in the TC_CMR which resets and restarts the timer when the desired count value is reached. The wave bit is set to put the timer into waveform mode. The bit values written to the register can be seen in **figure 14.1**. These bit values are equivalent to the hex number 0x0000C004 that is written to the register in our C program.

```
TC0_RC = 0x0000002F;
```

This register in known as register C. It contains the value that TC0 will count up to in order to generate the required timing value – in our case 1 millisecond (1ms). This register is 32 bits wide, but only the bottom two bytes of this register can be used, effectively making the value that can be written to it 16 bits.

Inside the **while(1)** loop, the current number of seconds that is stored in the **time** variable is sent out of the serial port for display on HyperTerminal. This value was initialised to 0 at the start of the program. The **TimeDelay()** function is then called and passed a value of 1000. This will cause the **TimeDelay()** function to count 1000 milliseconds (1 second) before it returns. The **time** variable is incremented in the **while(1)** loop after **TimeDelay()** returns which counts the second that has elapsed.

Inside the **TimeDelay()** function, TC0 is used to generate the time delay. The timer is first started with the following line:

```
TC0_CCR = 0x00000005;
```

This is the Timer Counter Channel Control Register (TC_CCR). The CLKEN bit in this register is set to enable the clock to the timer counter unit. The SWTRG bit is set which resets the counter and then starts it.

The **`while`** loop inside the **`TimeDelay()`** function will exit when the value passed to this function becomes 0. This value is decremented after every millisecond. The **`if`** statement is used to check when the timer has "triggered" which will result when the millisecond has been counted. The CPCS bit in the Status Register (TC_SR) is checked in the **`if`** statement. This bit will only be set when the timer counter has counted up to the value that we wrote to TC0_RC (the C register) at the beginning of the program.

14.2 The Analogue to Digital Converter (ADC)

The analogue to digital converter is described in chapter 36 of the AT91SAM7S datasheet. Before we can use the ADC, we need to solder a potentiometer to one of the ADC pins of the microcontroller. **Figure 14.2** shows the AT91SAM7S-EK evaluation board with a potentiometer attached to it so that a varying analogue voltage can be produced on one of the microcontroller's ADC pins.

Figure 14.2: Potentiometer Attached to the AT91SAM7S-EK Evaluation Board

The centre wiper of the potentiometer must be attached to one of the ADC pins of the microcontroller (AD4 to AD7), one of the end connections must be attached to 0V and the other to +3.3V. The AT91SAM7S-EK brings four analogue inputs to solder pads on the board. Two of these pins are from the set of four dedicated analogue inputs (AD4 and AD5), the other two pins are ADC pins that are multiplexed with I/O pins. The next program will use only the dedicated ADC pins, so be sure to connect your potentiometer to only AD4 to AD7 and not any of the multiplexed ADC inputs. If you are using a different evaluation board, it may already have a potentiometer on one of the dedicated pins or may have some other device that produces an analogue voltage on one of these pins. This will be fine to use with the program. The potentiometer will be used to change the voltage on an ADC pin between 0V and 3.3V. The next program, **analogue**, will use the ADC converter to read the voltage on the ADC pin, convert it to a digital number and send the number out of the DBGU serial port for displaying on HyperTerminal.

```
#include <stdio.h>
#include "at91sam7s.h"
#include "dbgu.h"

int main(void)
{
    unsigned int ADC_data[4];
    char str[50];
    int ch_num;
    /* setup the ADC */
    ADC_CR = 0x00000001;            /* reset the ADC */
    ADC_MR = 0x0F1F3F00;
    ADC_CHER = 0x000000F0;
    DBGUInit(DBGU_115200);
    ADC_CR = 0x00000002;            /* start the ADC conversion */
    while (1) {
        if (ADC_SR & 0x000000F0) {
            ADC_data[0] = ADC_CDR4;
            ADC_data[1] = ADC_CDR5;
            ADC_data[2] = ADC_CDR6;
            ADC_data[3] = ADC_CDR7;
            for (ch_num = 0; ch_num < 4; ch_num++) {
                sprintf(str, "ADC%d = %4d 0x%03x ",
                        ch_num + 4, ADC_data[ch_num], ADC_data[ch_num]);
                DBGUTxMsg(str);
            }
            ADC_CR = 0x00000002;        /* start the ADC conversion */
            DBGUTxMsg("\r");
        }
    }
}
```

This program starts the ADC and reads the values that it converts on channels AD4 to AD7. The values are sent out of the serial port and displayed on HyperTerminal as a decimal number and a hexadecimal number for each of the four channels.

The program works by first setting up the ADC:
The ADC is first reset by setting the SWRST bit in the ADC Control Register (ADC_CR) with the following line:

```
ADC_CR = 0x00000001;
```

The ADC Mode Register (ADC_MR) is set up next:

```
ADC_MR = 0x0F1F3F00;
```

This writes a value to the PRESCAL bit field in the ADC_MR which sets up the ADC clock rate. The STARTUP bit field which sets the start up time of the ADC. The SHTIM bit field sets up the sample and hold time of the ADC.

Writing to the ADC Channel Enable Register (ADC_CHER) enables the ADC channels AD4 to AD7.

The ADC is started just before the program enters the **`while(1)`** loop by writing to the ADC Control Register (ADC_CR) and setting the START bit in this register.

Inside the **`if`** statement in the **`while(1)`** loop, bits EOC4 to EOC7 in the ADC Status Register (ADC_SR) are checked to see if the ADC conversion is complete on ADC channels AD4 to AD7. The body of the **`if`** statement is only run if one of the channels has had its analogue value converted.

When all four channels have been converted, the converted values are found in the four ADC Channel Data Registers (ADC_CDR4 to ADC_CDR7. These values are read out of the registers and stored in the **`ADC_data`** array. A **`for`** loop is then used to print the conversion results on HyperTerminal via the serial port. ADC_CR is written to again to start the next ADC conversion.

14.3 Using the Timer and Interrupt

Up until now we have been "polling" hardware - continually checking it in our program to see if some condition has occurred. If you look at the **`timer`** program at the beginning of this chapter, you will see one of the disadvantages of polling hardware. When the **`TimeDelay()`** function is called it uses up all of the microcontroller's processing time to check (poll) the timer. If this function was to be used in a more useful program, the rest of the embedded system would not be able to respond to user button presses or other hardware that may need to be serviced while this function was running. Interrupts on the microcontroller are available to solve this problem.

Interrupts can be configured on the microcontroller so that, for example, when using the timer, it can be started and left to run on its own while the main program continues to run. When the timer reaches the time-out value e.g. after 1ms, an interrupt will be triggered. This interrupts the current flow of the main program and a function called an Interrupt Service Routine (ISR) is run in order to service the timer. When the interrupt occurs, the current state of the microcontroller is saved, a jump occurs to the ISR function where code is run that will respond to the timer. When the ISR function has finished running, a jump occurs back to where the main program left off and the saved state of the microcontroller is restored. The program carries on running normally after this until another interrupt occurs.

The **`timer_interrupt`** program uses an interrupt that is generated by the timer to increment a global variable. This program must be loaded into Flash memory in order for the interrupt vector table to be placed correctly in memory. The interrupt vector table is found in the **startup.s** assembly language file and is used to direct program flow to the correct function (interrupt service routine) when an interrupt occurs. When you create this program, use the Flash interrupt template files (refer to appendix B) and then use SAM-BA to load the program to Flash memory as we did with our first embedded program. You will need to restore the SAM-BA boot loader to Flash memory when you want to load a new program to Flash or SRAM after this.

```
#include <stdio.h>
#include "at91sam7s.h"
#include "dbgu.h"

/* global free-running count (counting in seconds) */
unsigned int g_count_s = 0;
/* TC0 ISR (interrupt service routine) */
void SysTimerISR(void) __attribute__ ((interrupt));

int main(void)
{
    char str[50];

    AIC_IDCR = 0xFFFFFFFF;
    DBGUInit(DBGU_115200);
    /* setup TC0 interrupt and ISR */
    /* configure the Advanced Interrupt Controller (AIC) */
    /* edge triggered, highest priority interrupt  */
    AIC_SMR12 = 0x00000027;
    /* address of TC0 ISR to source vector reg. */
    AIC_SVR12 = (unsigned long)SysTimerISR;
    AIC_IECR = 0x00001000;        /* enable TC0 interrupt */
    /* enable the clock of the TC0 module through the PMC */
    PMC_PCER = 0x00001000;

    /* setup TC0 registers */
    TC0_CMR = 0x0000C004;         /* set mode, clocksource */
    TC0_RC = 0x0000B798;          /* RC Register value for 1s */
    TC0_IER = 0x00000010;         /* enable RC compare interrupt */
    TC0_CCR = 0x00000005;         /* start the timer */

    while (1) {
        sprintf(str, "%d\r", g_count_s);
        DBGUTxMsg(str);
    }
}

/* interrupt service routine (ISR) */
void SysTimerISR(void)
{
    unsigned int tc_status;       /* stores source of TC0 interrupt */

    tc_status = TC0_SR;           /* clear the TC0 interrupt */
    g_count_s++;                  /* increment global counter */
    AIC_EOICR = 0x00000000;       /* acknowledge interrupt */
}
```

This program sends a count in seconds out of the serial port to be displayed on HyperTerminal. The timer is set up and started as before, but this time it is set to trigger an interrupt every second. When the timer times out and triggers the interrupt, the code that is running in the **`while(1)`** loop is "interrupted" and the **`SysTimerISR()`** interrupt service routine is run. This ISR function simply increments the global variable **`g_count_s`** by one. When the ISR function returns, the code in the **`while(1)`** loop continues running where it left off at the interrupt. This code continually displays the value of the **`g_count_s`** global variable which is a count of elapsed time in seconds.

The first thing to take note of in this program is the function prototype for the interrupt service routine function:

```
void SysTimerISR(void) __attribute__ ((interrupt));
```

This function prototype uses some non-C-standard extensions which are specific to the GNU C compiler that tell the compiler that this function is an interrupt service routine. This enables the correct code to be generated when the ISR is called and returned from.

After entry into **`main()`**, all interrupts are disabled by setting all of the corresponding bits in the AIC_IDCR (Interrupt Disable Command Register). This register is part of the Advanced Interrupt Controller (AIC) which is a piece of hardware on the AT91SAM7S microcontroller that enables interrupt vectoring to be handled by hardware rather than software. Vectoring refers to calling the correct interrupt service routine when an interrupt occurs, as more than one interrupt can be set up and used. To find out more about the AIC, refer to chapter 23 in the AT91SAM7S datasheet.

The AIC_SMR (AIC Source Mode Register) is used to set up interrupt priority and interrupt source type. The priority only comes into play when more than one interrupt is enabled and is used to determine which interrupt is run first should two or more interrupts occur simultaneously. You will notice in the source code that AIC_SMR12 is used to set up the interrupt of TC0. This is because there are a number of sources of interrupts and therefore a number of AIC_SMR registers; one for each source. Table 10-1 in chapter 10 of the AT91SAM7S datasheet shows the peripheral ID of each piece of hardware on the AT91SAM7S microcontrollers. TC0 has a peripheral ID of 12 and this indicates that we must use AIC_SMR12 when setting up an interrupt for this hardware peripheral.

The AIC_SVR12 (AIC Source Vector Register 12) is loaded with the address of the interrupt service routine that is to be run when the TC0 interrupt triggers. The following line in the source code does this:

```
AIC_SVR12 = (unsigned long)SysTimerISR;
```

The name of the function, **`SysTimerISR`**, is cast to an unsigned long integer which effectively returns the address of this function in memory. This address is loaded into the AIC_SVR register that corresponds to TC0's peripheral ID, i.e. AIC_SVR12.

The AIC_IECR (AIC Interrupt Enable Command Register) is used to enable the TC0 interrupt by setting the bit that corresponds to the peripheral ID of TC0 i.e. bit 12 (PID12) in this register.

Timer Counter 0 (TC0) is set up as before, except that now we are using a value in TC0_RC that will cause the timer to trigger every second. TC0_IER (TC0 Interrupt Enable Register) is used to enable an interrupt to occur when the timer triggers. The timer is then started before the code in the **`while(1)`** loop is executed.

In order for the interrupt to work correctly, a line of code in the **`startup.s`** file was modified to make use of the AIC when an interrupt occurs. When the interrupt does occur, the address of the ISR, **`SysTimerISR()`**, is produced by the AIC as the vector address. This causes the **`SysTimerISR()`** function to be run. This function is not called by the program at all, and is only called in the event of a TC0 interrupt being generated.

14.4 The Watchdog Timer

The watchdog timer is a peripheral device on the AT91SAM7S microcontroller and can be seen in the microcontroller's memory map. The AT91SAM7S datasheet describes the watchdog timer thus:

The Watchdog Timer can be used to prevent system lock-up if the software becomes trapped in a deadlock.

This means that should the software running on the microcontroller "hang up" or crash, the watchdog timer will reset the microcontroller as if an operator had pressed the reset button on the system.

The watchdog timer is enabled automatically when the microcontroller is powered up. The timer inside this watchdog timer is setup up by default to count 16 seconds and when the 16 seconds are up, it resets the microcontroller. The program running on the microcontroller is required to periodically restart the watchdog timer so that it starts counting down the 16 seconds without resetting the microcontroller, the idea being that if the program crashes, it will not be able to restart the watchdog timer and the watchdog will therefore reset the microcontroller to get the software running again. The time out value of the watchdog can be changed from the maximum value of 16 seconds to a smaller timeout value by programming one of the watchdog timer registers.

Information on programming the watchdog timer can be found in chapter 16 of the AT91SAM7S datasheet. The datasheet states that the Mode Register (WDT_MR) can only be written to once, after this a write to this register will have no effect on the watchdog timer unless the microcontroller is reset. We have been using SAM-BA to load our programs to SRAM, which means that the SAM-BA boot code runs on the microcontroller before we load our own program. Because the watchdog timer is enabled by default, SAM-BA would either set up the watchdog or disable it which means it would have to write to WDT_MR. This means that this register has already been written to once when our program starts, so if we try to set up the watchdog timer by writing to WDT_MR it won't work. Our program must therefore be compiled for and run from

Flash memory so that it is the only program that is run when the microcontroller is started up.

The **watchdog** program that follows is created from the **Flash_template** files – see appendix B for details. These template files are the same as the **RAM_template2** files except that they compile and link for Flash memory instead of RAM.

The **startup.s** file must be modified as it normally disables the watchdog timer. The three lines of assembly language code that are used to disable the watchdog timer are "commented out" which means that they are turned into a comment by using the C style open and close comment around them. This is necessary as the register used to disable the watchdog is the write once WDT_MR. The **watchdog** program does not actually write to the WDT_MR, but leaves the watchdog timer on to reset the processor after 16 seconds – its default.

The lines commented out in the **startup.s** file are as follows:

```
/* disable the Watchdog Timer (WDT) */
/*  LDR     R0, =WDT_BASE
    LDR     R1, =WDT_MR_Val
    STR     R1, [R0, #WDT_MR] */
```

Be sure to comment these lines out when you create this program from the template files.

Here is the **watchdog main.c** source code:

```
#include <stdio.h>
#include "at91sam7s.h"
#include "dbgu.h"

#define WDT_RESET_VAL   0xA5000001

void Delay(void);

int main(void)
{
    int count = 0;
    char str[50];

    DBGUInit(DBGU_115200);
    DBGUTxMsg("\r\nProgram starts from beginning...\r\n");
    DBGUTxMsg("Entering loop that restarts watchdog.\r\n\r\n");
    /* watchdog timer is restarted in this loop */
    while (1) {
        Delay();
        sprintf(str, "\rcount = %4d", count);
        DBGUTxMsg(str);
        count++;
        WDT_CR = WDT_RESET_VAL;      /* restart the watchdog timer */
        if (count > 50) {
            count = 0;
```

```
            break;
        }
    }
    DBGUTxMsg("\r\n\r\nNow not restarting watchdog.\r\n");
    DBGUTxMsg("Expect the microcontroller to be reset...\r\n\r\n");
    /* watchdog timer is not restarted in this loop */
    while (1) {
        Delay();
        sprintf(str, "\rcount = %4d", count);
        DBGUTxMsg(str);
        count++;
    }
}
```

The **`Delay()`** function is not included in this listing as you have used it before. Load the program to Flash memory using SAM-BA. Connect to HyperTerminal to see the output from this program.

The program starts by displaying a message to show that the program has started running from the beginning. The watchdog timer will be enabled to its default 16 second timeout because we have prevented it from being disabled in the assembly language start-up file.

A first **`while(1)`** loop is entered that displays a count periodically on HyperTerminal by running the delay routine. The watchdog timer is restarted periodically by writing to the WDT_CR. After a count of 50 is reached, the first **`while(1)`** loop is broken out of, a message is displayed on HyperTerminal to show this and a second **`while(1)`** loop is entered.

In the second **`while(1)`** loop, the same count is displayed on HyperTerminal, but this time the watchdog timer is not restarted. This means that the watchdog will time out and it will then reset the microcontroller. When it does, you will see the same entry message that was displayed when the program was first started and the counting will start from the beginning again just as if the microcontroller had been powered up or manually reset.

14.5 Summary

- In order to program an embedded system in any language, knowledge of the hardware being programmed is required.

- The timer counter in the AT91SAM7S microcontroller can be used to generate accurate timing intervals.

- An analogue to digital converter (ADC) is used to read external analogue values and convert them into digital numbers.

- Polling hardware refers to continually checking the hardware in a program to see if an event has occurred.

- Interrupts can be used in a C program to service hardware while the main program is running without having to call the servicing function in the code; it is called automatically when the interrupt occurs.

- The watchdog timer on a microcontroller can be used to recover from a software crash by resetting the microcontroller should this occur.

15 Wrapping Up

What you will do in this chapter:

- Use structures to group different data types together.
- Use pointers to access structures.
- Learn about unions.
- See how the enumerated data type works.
- Look at miscellaneous topics in C to complete the coverage of the C language.

15.1 Structures

Arrays allow us to store and access more than one variable of the same type. Structures can be used to group data of different types. For example a structure can contain a float, int and a char, but an array could only be of one of these types.

The **sample_struct** program uses a structure to group a timestamp value and sample data value together in order to demonstrate the use of structures in C.

```
#include <stdio.h>
#include "dbgu.h"

int main(void)
{
    char str[50];
    struct sample {           /* declare a data type structure */
        int time;             /* integer member of structure   */
        float data;           /* float member of structure     */
    };
    struct sample samp1;      /* define a structure of type sample */
    samp1.time = 20;          /* assign values to struct members   */
    samp1.data = 56.23;
    DBGUInit(DBGU_115200);
    sprintf(str, "\r\nTime in seconds is: %d\r\n", samp1.time);
    DBGUTxMsg(str);
    sprintf(str, "Sampled data is: %.2f", samp1.data);
    DBGUTxMsg(str);
    while (1);
}
```

A structure can contain any number of different data types and the compiler therefore does not know what you are going to put in a structure, so the structure must first be declared. In the **sample_struct** program, this is done with the following statement:

```
struct sample {
    int time;
    float data;
};
```

This declares a new data type called **struct sample** which has two member variables. The name **sample** is known as a tag and is a type name – i.e. we have declared a new data type that can now be used in our program. **Figure 15.1** shows the details of a C structure declaration.

Just like we can use the keyword **int** to define an integer variable in a program, we can now use the new data type **struct sample** to define a new structure of this type. This is what the next statement in the program does:

```
struct sample samp1;
```

The declaration of the new structure type did not allocate any memory for the structure, it just tells the compiler of the new data type that we are going to use in our program. The above statement now creates a new variable called **samp1** of type **struct sample**. This allocates memory for the structure and the variable name for this structure is now **samp1**. We could declare another variable in the program of type **struct sample** as follows:

```
struct sample samp2;
```

This will define another structure in the program of type **struct sample** with its own name. In other words a second structure of the same type would exist in memory that can be accessed by using its name – **samp2**, just like we could define two integers or other variable types in a program:

```
int val1;
int val2;

struct sample samp1;
struct sample samp2;
```

Each member variable of the structure can be accessed with the dot (**.**) operator as we can see in the program:

```
samp1.time = 20;
samp1.data = 56.23;
```

In each case, the structure name followed by the dot operator and then the name of the member variable is used to access the member variable of the structure.

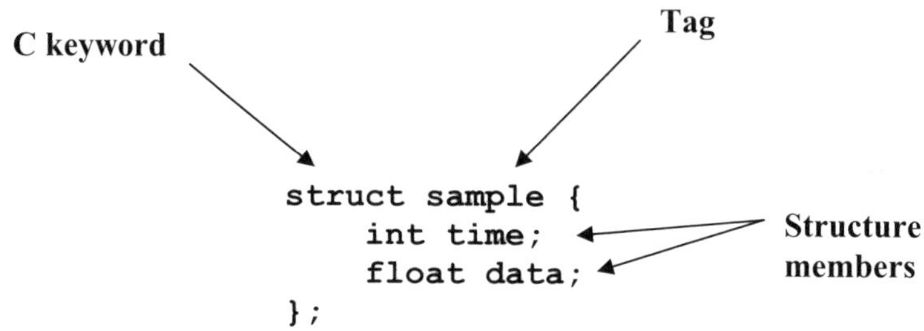

Figure 15.1: A C Structure Declaration

Structures, like arrays can be initialised with data values. The **init_struct** program initialises three structures that each contain a pattern to display on the LEDs and the duration to display it for.

```
#include "at91sam7s.h"

void InitHardware(void);
void TimeDelay(unsigned int ms_delay);

int main(void)
{
    struct pattern_time {
        unsigned char pattern;
        unsigned int time_ms;
    };
    struct pattern_time pt1 = {0x05, 2000};
    struct pattern_time pt2 = {0x0A, 5000};
    struct pattern_time pt3 = {0x03, 3000};

    InitHardware();

    while (1) {
        PIO_ODSR = ~pt1.pattern;
        TimeDelay(pt1.time_ms);
        PIO_ODSR = ~pt2.pattern;
        TimeDelay(pt2.time_ms);
        PIO_ODSR = ~pt3.pattern;
        TimeDelay(pt3.time_ms);
    }
}

void InitHardware(void)
{
    /* configure PIO lines for LEDs on PA0 to PA3 */
    PIO_PER = 0x0000000F;       /* enable uC pins as PIO */
    PIO_OER = 0x0000000F;       /* enable uC pins as output */
    PIO_OWER = 0x0000000F;      /* allow direct writing to port bits */
    PMC_PCER = 0x00001000;      /* enable TC0 clock */
    TC0_CMR = 0x0000C004;       /* set mode, clocksource */
    TC0_RC = 0x0000002F;        /* RC Register value for 1ms */
}
```

```
void TimeDelay(unsigned int ms_delay)
{
    TC0_CCR = 0x00000005;         /* start the timer */
    while (ms_delay) {
        if (TC0_SR & 0x00000010)
            ms_delay--;
    }
}
```

This program calls the function **`InitHardware()`** to initialise the hardware that is used in the program. This makes the code in **`main()`** less cluttered and easier to read.

A structure is declared that contains two member variables that will be used to store a pattern to be displayed on the LEDs as well as the duration in milliseconds to display the pattern.

Three structures are then defined and initialised using the structure that was declared. This makes space in memory for three structures named pt1 to pt3 and assigns values to their member variables. In the structure, the **`pattern`** member variable occurs first and the **`time_ms`** member variable occurs second. When initialisation takes place, the first value in the braces is assigned to the first member variable – **`pattern`** and the second value in the braces is assigned to the second member variable – **`time_ms`**.

15.1.1 Pointers to Structures

The next example, **`point_struct`**, uses a pointer to a structure in order to show the syntax used to access structure members with a pointer.

```
#include <stdio.h>
#include "dbgu.h"

int main(void)
{
    char str[50];
    struct time {
        unsigned char hour;
        unsigned char minute;
        unsigned char second;
    };
    struct time cur_time = {12, 34, 27};    /* structure variable */
    struct time *t_ptr;                     /* pointer to struct */
    t_ptr = &cur_time;
    DBGUInit(DBGU_115200);
    sprintf(str, "Time is %02d:%02d:%02d\r\n",
            t_ptr->hour, t_ptr->minute, t_ptr->second);
    DBGUTxMsg(str);
    while (1);
}
```

This program first declares a structure of type **`struct time`**. A structure of type **`struct time`** called **`cur_time`** is then defined and initialised. Next a pointer to a structure of type **struct time** is defined called **t_ptr**. This makes space in memory to store the address of a structure of this particular type.

The line: **t_ptr = &cur_time;** assigns the address of the **cur_time** structure to the structure pointer **t_ptr**. When the contents of the structure is accessed by the pointer in the **sprintf()** function, the arrow operator (**->**) is used to access each member variable. The arrow operator is always used to access member variables of a structure when we are using a pointer to a structure. The dot operator is always used to access the member variables of a structure when we are not using a pointer to a structure, but are accessing the structure directly.

15.2 Unions

A union in C uses the same syntax as a structure, except that the keyword **union** is used instead of **struct** when declaring it. The members of a union share the same space in memory which allows us to access the single space in memory as one type of variable or another. The **union** program will demonstrate.

```
#include <stdio.h>
#include "dbgu.h"

int main(void)
{
    char str[50];
    union unnum {
        float flt;
        int   dec;
    };
    union unnum my_union;
    DBGUInit(DBGU_115200);
    my_union.flt = 23.7;
    sprintf(str, "\r\nnum as float is %.2f\r\n", my_union.flt);
    DBGUTxMsg(str);
    sprintf(str, "num as int is %d\r\n", my_union.dec);
    DBGUTxMsg(str);

    my_union.dec = 100;
    sprintf(str, "\r\nnum as float is %.2f\r\n", my_union.flt);
    DBGUTxMsg(str);
    sprintf(str, "num as int is %d\r\n", my_union.dec);
    DBGUTxMsg(str);
    while (1);
}
```

A union called **unnum** is declared in this program. A union called **my_union** is then defined. The union contains a floating point member variable as well as an integer member variable. On the ARM microcontroller, both of these variables occupy four bytes of memory. When they occur in a union, they both occupy the same four bytes of memory, so the same four bytes of memory can be viewed as an integer or a floating point number.

In the program the floating point member variable is first initialised with the value 23.7. This value is then displayed on HyperTerminal as a floating point number and then as an integer. To display the number as a floating point number, we access the floating point member variable **flt**. To display this number (4 byte memory location) as an integer, we access the integer member variable **dec**. This results in a value of 1102944666 being displayed because the two numbers are in different formats i.e. the floating point number contains an integer part and a fractional part in the four byte memory location and this gives us a meaningless number when we view these four bytes as a decimal number.

To prove that we are looking at the same four bytes in memory, the integer member variable is initialised to a value of 100. This causes the floating point member variable to change to 0.00 when printed to HyperTerminal.

A more practical program follows called **multibyte** where we use a union to access an integer as a 32-bit integer or as individual bytes. This could be used in a program to access a 32 bit port as 32 bits or as four single bytes.

```
#include <stdio.h>
#include "dbgu.h"

int main(void)
{
    char str[50];
    int count;
    union wrd_byte{
        unsigned int word;
        unsigned char byte[4];
    } port;

    DBGUInit(DBGU_115200);
    port.word = 0xFF5501AA;
    sprintf(str, "\r\nWord is: 0x%08X\r\n", port.word);
    DBGUTxMsg(str);
    for (count = 0; count < 4; count++) {
        sprintf(str, "Byte%d: 0x%02X\r\n", count, port.byte[count]);
        DBGUTxMsg(str);
    }
    while (1);
}
```

In this program a union is declared and a union with the name **port** is defined in one step by putting the name of the union to be defined between the closing brace and the terminating semicolon of the union declaration.

The union consists of two member variables: a four byte integer called **word** and a four byte array called **byte**. In the program the **word** member variable is initialised to a value of 0xFF5501AA and then printed on HyperTerminal. A **for** loop is then used to print the same number as four individual bytes by accessing each byte of the **byte** array. Of course because the **byte** and **word** member variables share the same four bytes in memory, the value that **word** was initialised to is displayed on HyperTerminal in four parts starting with the least significant byte in **word**. **Figure 15.2** shows what this looks like in memory.

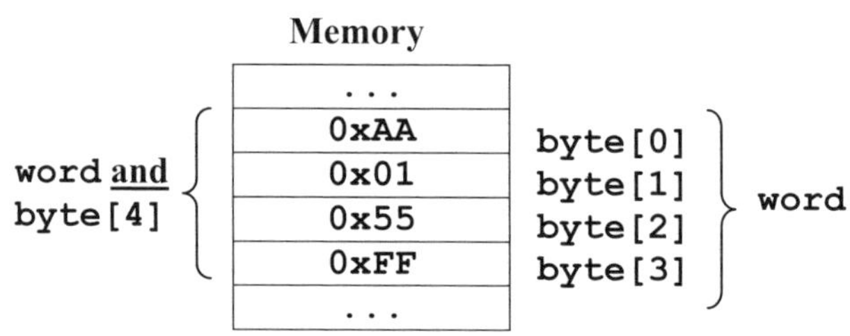


```
word = 0xFF5501AA
byte[0] = 0xAA
byte[1] = 0x01
byte[2] = 0x55
byte[3] = 0xFF
```

Figure 15.2: A union in Memory

15.3 Enumerated Type

The enumerated data type in C is an integer data type that can be used to make a program easier to read and understand. The enumerated data type allows you to create your own data type and specify the values that can be assigned to it. If an attempt is made to assign a value other than you have specified, the compiler will generate an error.

The next program uses the embedded system as a controller for a dough mixing machine. Three LEDs on port pins PA0 to PA2 represent three solenoids that are used to control the machine. The switch on pin PA15 is used by the machine operator to start the machine. The hardware is configured as follows:

Outputs:
PA0 = water pump solenoid – adds water to the mixing bowl when activated
PA1 = flour hopper solenoid – adds flour to the mixing bowl when activated
PA2 = mixer – starts and runs the mixer when activated

Input:
PA15 = start button – pressed by the machine operator to make a batch of dough

Run the program and press the button on PA15 to start the dough mixing procedure. Each of the LEDs (solenoids) will switch on in sequence so that water is added to the bowl for

4 seconds, flour is added for 2 seconds and the dough is mixed for 5 seconds. After this the machine goes back into standby mode and waits for the operator to press the start button again to make another batch of dough. Here is the **enum_seq** program used to demonstrate enumerated data types in C:

```
#include "at91sam7s.h"
#include "dbgu.h"

void InitHardware(void);
void TimeDelay(unsigned int ms_delay);

int main(void)
{
    enum machine_state { standby, pump_water, add_flour, mixing };
    enum machine_state state = standby;

    InitHardware();
    while (1) {
        switch (state) {
            case standby:
                /* switch all solenoids off */
                PIO_ODSR = ~0x00;
                /* check for start button pressed */
                if (!(PIO_PDSR & 0x00008000)) {
                    state = pump_water;
                }
            break;

            case pump_water:
                /* switch water pump solenoid on for 4s */
                PIO_ODSR = ~0x01;
                TimeDelay(4000);
                state = add_flour;
            break;

            case add_flour:
                /* switch flour hopper solenoid on for 2s */
                PIO_ODSR = ~0x02;
                TimeDelay(2000);
                state = mixing;
            break;

            case mixing:
                /* mix dough for 5s */
                PIO_ODSR = ~0x04;
                TimeDelay(5000);
                state = standby;
            break;

            default:
            break;
        }
    }
}
```

The **HardwareInit()** and **TimeDelay()** functions have been left out of the listing for clarity. You have used both of these functions in previous programs with the exception that **HardwareInit()** now enables PA15 as an input. Refer to the source code for this project for the code in these functions.

Inside **main()** an enumerated data type is declared:

```
enum machine_state {
         standby, pump_water, add_flour, mixing };
```

The **enum** keyword tells the compiler that this is to be an enumerated data type. The tag **machine_state** gives it a name so that we can use it in our program. The four names in between braces are the four possible values that an enumerated data type of this kind can take on. These names are actually integer values that start from 0 and count up, so **standby** is the value 0, **pump_water** is the value 1, **add_flour** is the value 2, etc. These are not variable names, but are actual integers that we refer to by these meaningful names.

The next line in the program creates a variable of type **enum machine_state** called **state** and assigns it the value of **standby**:

```
enum machine_state state = standby;
```

We now have a variable called **state** that can take on only one of the values that are part of the enumerated type i.e. **standby**, **pump_water** etc.

Next, the hardware that we are going to use in the program is initialised by calling the **InitHardware()** function. This sets up PA15 as an input, PA0 to PA3 as outputs (although we don't use PA3 in this program) and timer counter TC0 is set up to count in milliseconds.

A **switch** statement is used to sequence the machine. The **switch** statement used in this way is known as a "state machine" as it controls the state of the program. When the **switch** statement is run, it checks the value of the enumerated variable **state**. Because we initialised this variable to the value of **standby**, the first **case** statement will be continually run as it is inside the **while(1)** loop. This **case** statement is used to check if the start button is pressed by the machine operator. It also switches all of the solenoid valves (LEDs) off. Should the operator press the start button that is wired to pin PA15, the value of the enumerated data type **state** is changed to the value of **pump_water** inside the **if** statement. This results in the next **case** statement (**pump_water**) being run the next time that the **switch** statement is executed.

The **pump_water case** switches on the water pump solenoid for four seconds and then changes the state of the state machine to **add_flour**. The **add_flour** and **mixing** states operate in a similar manner, each switching off the previous solenoid valve and switching on a new one, then waiting for a specified time period before changing the state of the state machine. When the **mixing** state has been completed, the state is changed back to **standby** and the push button switch is again monitored for the operator to start a new batch of dough.

Note how the enumerated data type makes the program easier to understand as it uses easy to understand names instead of numbers to represent the four states of the machine.

15.4 The typedef Declarator

The **typedef** keyword enables us to give another name to a C data type. This can be used to make a program easier to read. The following lines show how to use **typedef** to give a different name to a C data type:

```
typedef unsigned char BYTE;
typedef unsigned char UCHAR;
```

Either **BYTE** or **UCHAR** can now be used in a program wherever we would have used **unsigned char** – this has effectively made a new C data type. Using the name **BYTE** is more descriptive especially when programming hardware. **UCHAR** is a shorter way of writing **unsigned char** and can save some typing and space on a line especially when a function takes a number of arguments that are of this type. In the program we can now define variables of this type:

```
BYTE port1_val;
BYTE port2_val;
```

And is that same as:

```
unsigned char port1_val;
unsigned char port2_val;
```

typedef can also be very useful when used to create a new structure. The **typedef** declaration can be made in a header file and every source file that the header file is included in can make use of the new data type. As an example:

```
typedef struct {
    unsigned char hours;
    unsigned char minutes;
    unsigned char seconds;
} time_struct;
```

Notice that we do not even need a tag name in the **typedef** declaration. This structure can now easily be used in any program as follows:

```
time_struct curr_time = { 2, 32, 56 };
time_struct alarm_time = { 3, 30, 00 };
```

We have effectively created a new data type called **time_struct** and can now create variables of this type by using its type name.

15.5 Storage Class Specifiers

There are four storage class specifiers in C of which you have already seen the **static** specifier. The other three are **auto**, **register** and **extern**. Only one storage class specifier may appear in a declaration. The meaning of each of these specifiers is described below:

auto – This keyword is actually outdated as any variable that is declared in a function is an automatic variable. When a variable is declared with the **auto** specifier it will have automatic storage duration. This means that when a function is called, the automatic variable will be created (it will be given space in memory) and when the function exits, the variable will be destroyed (the memory will be freed for use by other parts of the program). This happens by default anyway. E.g:

```
auto int val;
```

extern – Functions and variables declared with the **extern** keyword can be used anywhere in the program. They will have "global" visibility. E.g:

```
extern int val;
```

register – This keyword is a suggestion to the compiler that you want to store a variable in an internal CPU register to make access to it faster. This is not however guaranteed to happen. You can not use the address operator with variables that are declared with this specifier. E.g:

```
register float fast_flt;
```

static – Variables that are declared **static** will retain their value even when they go out of scope e.g. when the function in which they exist exits, they will not be destroyed. E.g:

```
static unsigned int count;
```

An example of the use of the **extern** keyword is if our C program consists of more than one file and we want to be able to use the global variable that was created in one file in the second file, we would use the **extern** keyword in the second file. For example:

In **main.c** we create a global variable (a variable defined outside of a function):

```
int globvar = 2;
```

We now want to use this global variable in another file in this program – **comms.c**. In **comms.c** we add the following at the top of the file or in a header file that is included in **comms.c**:

```
extern int globvar;
```

We are now free to use **globvar** in the **comms.c** file.

15.6 Type Qualifiers

C has two type qualifiers namely **const** and **volatile**.

const – a variable that is qualified with the **const** keyword is constant and it can not be changed by the program. In embedded programming this qualifier is used to force data to be put in the non-volatile memory of the system e.g. into Flash memory of the AT91SAM7S microcontroller. E.g.:

```
const int ref = 3;
```

volatile – we have already used the **volatile** keyword with pointers to registers on the microcontroller. This keyword forces the program to reread the value that is qualified as volatile whenever it is used as it may be modified by an external process or event. Think of an external register that the hardware device modifies and changes without the program changing it. For example the receive register of a serial port will contain a new value when data is received on the serial port. The program has nothing to do with changing this register's value, the serial port hardware does this.

15.7 The goto Statement

Use of the **goto** statement is considered bad programming practice by most C programmers. The **goto** statement can make programs hard to understand and debug because it can cause program flow to jump anywhere in a program. The use of the **goto** statement is unnecessary in a program and there will always be a better way to write a program without this statement by using a loop or **if** statement. How to use the **goto** statement is included here as you may come across it in someone else's code and will need to understand how it works.

The **goto** statement is always used in conjunction with a label. An example will illustrate:

```
if (time == alarm_time)
    goto ring_alarm;
else
    CountSeconds();

ring_alarm:
    AlarmOn();
    TimeDelay(5000);
    AlarmOff();
```

The **goto** statement will cause a jump to the **ring_alarm** label and the code under this label will be run.

15.8 A List of All C Keywords

Table 15.1 contains a list of all 32 C keywords as a summary of what we have covered.

auto	double	int	struct
break	else	long	switch
case	enum	register	typedef
char	extern	return	union
const	float	short	unsigned
continue	for	signed	void
default	goto	sizeof	volatile
do	if	static	while

Table 15.1: All C Keywords

15.9 More Preprocessor Directives

We have used the **#include** and **#define** preprocessor directives. There are other preprocessor directives that can be used for "conditional compilation". The **#ifdef** directive can be used to compile certain parts of a program and not others when used in combination with the **#define** and **#endif** directives.

An example of the use of conditional compilation is if we want to write a C program that can run on say the AT91SAM7S microcontroller and the LPC2148 microcontroller. We would want to compile the program to use the AT91SAM7S hardware to run on this microcontroller, but also want to be able to compile the same program to use the LPC2148 hardware. The example code that follows will show how to do this.

```
#define ATMEL

int main(void)
{
#ifdef ATMEL
    PIO_CODR = 0x00000001;      /* switch LED on */
#else
    IOCLR0 = 0x00800000;        /* switch LED on */
#endif
}
```

In this example PIO_CODR is the register that is written to on the AT91SAM7S to clear an external microcontroller pin. IOCLR0 is the register written to on the LPC2148 to do the same thing.

When the program is compiled with **ATMEL** defined as above, the code under the **#ifdef ATMEL** directive will be included in the program by the pre-processor and the code under the **#else** directive will be excluded. When the **#define ATMEL** directive is removed from the file, only the code under the **#else** directive will be compiled. The **#endif** directive is used to show that the **#ifdef** and **#else** directives end here – like the closing brace of a construct in C.

After the pre-processor has run, the compiler will see the following if **ATMEL** is defined:

```
int main(void)
{
    PIO_CODR = 0x00000001;
}
```

This code has been conditionally compiled to run on the AT91SAM7S microcontroller.

After the pre-processor has run, the compiler will see the following if **ATMEL** is not defined:

```
int main(void)
{
    IOCLR0 = 0x00800000;
}
```

This code has been conditionally compiled to run on the LPC2148 microcontroller.

15.10 Debugging

Debugging your embedded ARM system is well explained in the tutorial by James P. Lynch that has already referred to in this book. You will need to refer to the YAGARTO website on how to update Eclipse to install the Zylin Embedded CDT plug-in. This plug-in allows Eclipse to interface to an external JTAG debug device so that you can single-step debug embedded C programs. At the time of writing, the tutorial had not been updated to include installation of the plug-in, so the software installation instructions are a bit outdated – be sure to read the updated instructions on the YAGARTO website. Previously the plug-in was included with Eclipse and Eclipse was downloaded from the YAGARTO website. The rest of the tutorial, however, is still valid and if you have not read it yet, do so now.

15.11 Some Final Example Programs

Some more programs have been included in this section that will hopefully be of some use to you on the limited number of hardware devices that we have been using. Source listings are not included here; refer to appendix B for details.

15.11.1 Voltmeter

The purpose of the **Voltmeter** program is to use one of the ADC channels as a voltmeter that can measure voltage from 0V to 3.3V and display this voltage on

HyperTerminal. The program reads the value from the ADC and converts it into a meaningful floating point voltage that a user can read. An LED is switched on should the voltage on the ADC channel approach the maximum voltage value that the ADC can read.

On the AT91SAM7S-EK evaluation board, the reference voltage used with the ADC is the 3.3V supply voltage of the microcontroller. To find out what reference voltage is used on your board, look at the schematic diagram. On the AT91SAM7S microcontroller, pin 1 of the microcontroller is the ADVREF pin onto which the ADC reference voltage is supplied.

The 10-bit value that the channel data register (ADC_CDR) holds after an ADC conversion is changed to voltage as follows:

Each unit in the ADC_CDR has a voltage value of:
Reference Voltage ÷ 1023
= 3.3 ÷ 1023
= 0.003225806

Where 1023 is the highest value that the 10-bit ADC can count up to (2 to the power of 10 minus 1):
$\mathbf{2^{10} - 1 = 1023}$

To find the voltage value that the ADC_CDR represents, simply multiply the voltage that each unit in this register represents by the value in the register. E.g. if the register contains a value of 350, the voltage that this number represents is:
350 × 0.003225806
= 1.129V

To summarise, the formula for voltage calculation of the ADC value is:
ADC Voltage = (3.3 ÷ 1023) × ADC value

This is calculated in the **`GetVoltage()`** function and written in the C source code as:

```
voltage = (REF_VOLTAGE / 1023.0) * (float)ADC_data;
```

The line above this in the source code makes sure that the upper bits of the 32-bit ADC_CDR are cleared to zero, as only the 10 least significant bits in this register are valid. This is done with the bitwise **AND** operator:

```
ADC_data = ADC_CDR5 & 0x000003FF;
```

The program will switch an LED on if the calculated voltage comes close to the reference voltage in order to warn the user not to increase the voltage beyond this value which would damage the embedded system.

Values that may need to be changed in the program have been defined at the top of the source file. This makes it easy to compile the program to use a different ADC channel or to make the adjustment should the reference voltage on another board be different to the one used in the program. The threshold voltage at which the LED switches on can also be easily changed here.

15.11.2 Voltmeter with ADC Filtering

The **Voltmeter_filt** program does exactly the same as the previous program, but now calculates the average of a number of ADC samples in order to show a more stable voltage on HyperTerminal. The result of the averaging is a voltage in which the least significant digits do not flicker between values when displayed on HyperTerminal.

A buffer (array) is created to store 20 sequential ADC samples. The average of the 20 samples is calculated by adding these samples together and then dividing the result by 20. This buffer has been set up as a circular buffer. This means that the buffer will be filled with ADC values from the first element in the buffer to the last. When the last element has been filled, values will be written from the first element again.

In the function **GetVoltage()**, static variables have been used again for the variables that must not loose their values between function calls. The initialisation of the array that is used as a buffer is as follows:

```
static unsigned short ADC_buf[NUM_SAMPLES] = {0};
```

The notation here initialises each element of the array to zero without having to write a zero for each element in the array.

The size of the buffer has been defined at the top of the source file with the name **NUM_SAMPLES** so that it can easily be changed in the program. Try experimenting with different buffer sizes.

If you run this program and compare it to the previous one, you will notice that the displayed voltage is much more stable and that the LED does not flicker when the threshold voltage is approached.

15.11.3 Stopwatch

The **stopwatch** program uses the timer on the microcontroller to implement a stopwatch. A button on the embedded system is used to start the stopwatch and a second button is used to stop the stopwatch. The time is displayed on HyperTerminal.

A variable called **timer_off** is used as a flag in this program to prevent the timer from being started or stopped multiple times as the user holds the start or stop button in. This is also a way of "de-bouncing" the button. When a button is pressed, the switch contacts do not make a clean transition from off to on but rather "bounce" or switch on and off rapidly until they settle. This would have the effect of running the timer start code multiple times when the button is pressed and then again when it is released.

15.12 Summary

- Structures can be used to group data of different types together.
- The dot operator is used to access members of a structure.
- The arrow operator is used to access structure members when using a pointer to the structure.
- A union is similar to a structure but its member variables share the same memory space.
- The enumerated data type can be used to make a program easier to read and understand.
- The **`typedef`** declarator can be used to rename a data type or give a name to a new data type such as a user defined structure.
- Storage class specifiers (**`auto`**, **`extern`**, **`static`**, **`register`**) can be used to modify the way a program treats a variable.
- The **`const`** keyword specifies that a variable is constant and cannot be changed in the program. If an attempt is made to modify a constant variable, the compiler will generate an error. Constant variables will be put into the non-volatile memory of an embedded system by the linker.
- The **`volatile`** keyword is used wherever a register or variable can be changed by hardware that operates externally to the C program and can change the value of the register or variable without the program knowing about the change.
- The **`goto`** statement can make a program difficult to understand as it can cause a jump to anywhere in the program. It is considered good programming practice to not use the **`goto`** statement as there is always a better way of writing a program without this statement.

I hope that you have enjoyed using this book to learn the C programming language. Although it would have been nice to have programmed more embedded hardware devices, limited space in this book as well as an attempt to make it possible to use a number of different embedded systems with the book, prevented this. I will leave you to explore other hardware devices on your own.

Where to from here?
The key to becoming a proficient C programmer is to write programs as well as to read other peoples programs. Use the Internet to obtain source code that you can read and learn from. Write lots of your own programs and never be afraid to experiment.

Another good resource for embedded microcontroller programmers is the websites of microcontroller manufacturers. Manufacturers have application notes available that cover various topics including programming of various devices on the microcontrollers. The application notes on programming are usually accompanied by source code so be sure to download the code and see how it works.

Good luck! I hope that you spend many enjoyable hours programming embedded systems in C!

Appendix A: The ASCII Table

DEC	HEX	CHAR	DEC	HEX	CHAR	DEC	HEX	CHAR	DEC	HEX	CHAR
0	0	NULL	32	20	space	64	40	@	96	60	`
1	1	SOH	33	21	!	65	41	A	97	61	a
2	2	STX	34	22	"	66	42	B	98	62	b
3	3	ETX	35	23	#	67	43	C	99	63	c
4	4	EOT	36	24	$	68	44	D	100	64	d
5	5	ENQ	37	25	%	69	45	E	101	65	e
6	6	ACK	38	26	&	70	46	F	102	66	f
7	7	BEL	39	27	'	71	47	G	103	67	g
8	8	BS	40	28	(	72	48	H	104	68	h
9	9	HT	41	29	)	73	49	I	105	69	i
10	0A	LF	42	2A	*	74	4A	J	106	6A	j
11	0B	VT	43	2B	+	75	4B	K	107	6B	k
12	0C	FF	44	2C	,	76	4C	L	108	6C	l
13	0D	CR	45	2D	-	77	4D	M	109	6D	m
14	0E	SO	46	2E	.	78	4E	N	110	6E	n
15	0F	SI	47	2F	/	79	4F	O	111	6F	o
16	10	DLE	48	30	0	80	50	P	112	70	p
17	11	DC1	49	31	1	81	51	Q	113	71	q
18	12	DC2	50	32	2	82	52	R	114	72	r
19	13	DC3	51	33	3	83	53	S	115	73	s
20	14	DC4	52	34	4	84	54	T	116	74	t
21	15	NAK	53	35	5	85	55	U	117	75	u
22	16	SYN	54	36	6	86	56	V	118	76	v
23	17	ETB	55	37	7	87	57	W	119	77	w
24	18	CAN	56	38	8	88	58	X	120	78	x
25	19	EM	57	39	9	89	59	Y	121	79	y
26	1A	SUB	58	3A	:	90	5A	Z	122	7A	z
27	1B	ESC	59	3B	;	91	5B	[	123	7B	{
28	1C	FS	60	3C	<	92	5C	\	124	7C	|
29	1D	GS	61	3D	=	93	5D	]	125	7D	}
30	1E	RS	62	3E	>	94	5E	^	126	7E	~
31	1F	US	63	3F	?	95	5F	_	127	7F	delete

0 to 31 are unprintable control codes used for communications
32 to 127 are printable ASCII characters

DEC	HEX	CHAR	DEC	HEX	CHAR	DEC	HEX	CHAR	DEC	HEX	CHAR
128	80	Ç	160	A0	á	192	C0	└	224	E0	α
129	81	ü	161	A1	í	193	C1	┴	225	E1	ß

130	82	é	162	A2	ó	194	C2	┬	226	E2	Γ
131	83	â	163	A3	ú	195	C3	├	227	E3	π
132	84	ä	164	A4	ñ	196	C4	─	228	E4	Σ
133	85	à	165	A5	Ñ	197	C5	┼	229	E5	σ
134	86	å	166	A6	ª	198	C6	╞	230	E6	µ
135	87	ç	167	A7	º	199	C7	╟	231	E7	τ
136	88	ê	168	A8	¿	200	C8	╚	232	E8	Φ
137	89	ë	169	A9	⌐	201	C9	╔	233	E9	Θ
138	8A	è	170	AA	¬	202	CA	╩	234	EA	Ω
139	8B	ï	171	AB	½	203	CB	╦	235	EB	δ
140	8C	î	172	AC	¼	204	CC	╠	236	EC	∞
141	8D	ì	173	AD	¡	205	CD	═	237	ED	φ
142	8E	Ä	174	AE	«	206	CE	╬	238	EE	ε
143	8F	Å	175	AF	»	207	CF	╧	239	EF	∩
144	90	É	176	B0	░	208	D0	╨	240	F0	≡
145	91	æ	177	B1	▒	209	D1	╤	241	F1	±
146	92	Æ	178	B2	▓	210	D2	╥	242	F2	≥
147	93	ô	179	B3	│	211	D3	╙	243	F3	≤
148	94	ö	180	B4	┤	212	D4	╘	244	F4	⌠
149	95	ò	181	B5	╡	213	D5	╒	245	F5	⌡
150	96	û	182	B6	╢	214	D6	╓	246	F6	÷
151	97	ù	183	B7	╖	215	D7	╫	247	F7	≈
152	98	ÿ	184	B8	╕	216	D8	╪	248	F8	°
153	99	Ö	185	B9	╣	217	D9	┘	249	F9	∙
154	9A	Ü	186	BA	║	218	DA	┌	250	FA	·
155	9B	¢	187	BB	╗	219	DB	█	251	FB	√
156	9C	£	188	BC	╝	220	DC	▄	252	FC	^{n}
157	9D	¥	189	BD	╜	221	DD	▌	253	FD	^{2}
158	9E	Pts	190	BE	╛	222	DE	▐	254	FE	■
159	9F	ƒ	191	BF	┐	223	DF	▀	255	FF	

128 to 255 are extended ASCII characters

Appendix B: Source Code and Software

The source code and software used in this book is downloadable from the Elektor website:

www.elektor.com

On the front page of the Elektor website, click **Books** under **Products**. Click the **Microcontrollers** tab and then the **C Programming for Embedded Microcontrollers** link. Click the **ExtrasCprogramming.zip** link to download the source code and the DJGPP compiler. The `Source_Code.zip` file in the `ExtrasCprogramming.zip` file contains all source code used throughout the book. The directories mentioned in this section that contain the source code files are found in this zipped file.
A list of each source code file and software program used in each chapter follows:

PC programs Chapter 1 to Chapter 7

Chapter 1

Example Programs:
`Source_Code\PC\CH1\`
`hello.c`

Exercise Solutions:
`Source_Code\PC\Exercises\CH1\`
1. `ex1.c`
2. `ex2.c`

Chapter 2

Example Programs:
`Source_Code\PC\CH2\`
`sticky.c`
`number.c`
`var.c`
`add.c`
`input.c`
`inout.c`
`intdiv.c`
`fltdiv.c`
`charvar.c`
`ohm.c`
`fltdiv.c`
`getchit.c`

Exercise Solutions:
`Source_Code\PC\Exercises\CH2\`
1. a) `1a_resistance.c`
   b) `1b_voltage.c`
2. `2_var_test.c`

Chapter 3

Example Programs:
`Source_Code\PC\CH3\`
`compare.c`
`whatif.c`
`ifelse.c`
`fussy.c`
`multif.c`
`pumpctrl.c`

Exercise Solutions:
`Source_Code\PC\Exercises\CH3\`
1. `1_compare.c`
2. `2_relation.c`

Chapter 4

Example Programs:
`Source_Code\PC\CH4\`
`water.c`
`monit.c`
`guess.c`
`montemp.c`
`comment.c`
`style1.c`
`style2.c`
`style3.c`

Exercise Solutions:
`Source_Code\PC\Exercises\CH4\`
1. `subtract.c`
2. `pumpctrl_loop.c`

Chapter 5

Example Programs:
`Source_Code\PC\CH5\`
`countfunc.c`

passcount.c
twoargs.c
vararg.c
retval.c
addfunc.c
flashled.c
defdel.c
circle.c

splitprog.c
splitfunc.c

mkmain.c
mkgraph.c
mkgraph.h
makefile

<u>Exercise Solutions:</u>
Source_Code\PC\Exercises\CH5\

1. altflash.c
2. barctrl.c
3. barctrl2.c

Chapter 6

<u>Example Programs:</u>
Source_Code\PC\CH6\
printhex.c
printnum.c
convnum.c
padding.c

asciiread.c
textfile.txt

asciiwrite.c

<u>Exercise Solutions:</u>
Source_Code\PC\Exercises\CH6\

2. ex2.c
3. ex3.c

Chapter 7

<u>Example Programs:</u>
Source_Code\PC\CH7\

```
address.c
led_port.c
pointer.c
varsize.c
vardef.c
```

Embedded Programs Chapter 8 to Chapter 15

Template Files Used

Template files used for Chapter 9 and Chapter 10:
`Source_Code\Embedded\Templates\RAM_template\`

Template files used in Chapter 11 to Chapter 15:
`Source_Code\Embedded\Templates\RAM_template2\`
Use this template for all programs that run from SRAM and use the DBGU serial port drivers.

Except for the `timer_interrupt` program in chapter 14 that uses:
`Source_Code\Embedded\Templates\Flash_interrupt\`
Use this template for all programs that run from Flash, use interrupts and the DBGU serial port drivers.

And the watchdog program in chapter 14 that uses:
`Source_Code\Embedded\Templates\Flash_template\`
Use this template for all programs that run from Flash and use the DBGU serial port.

Chapter 8

Example Programs:
`Source_Code\Embedded\CH8\`
`LED_onoff`

Chapter 9

Example Programs:
`Source_Code\Embedded\CH9\`
`LED_two`
`switch_LED`

Exercise Solutions:
`Source_Code\Embedded\Exercises\CH9\`
1. `sw_blink`

Chapter 10

Example Programs:
`Source_Code\Embedded\CH10\`
`serial_tx`
`string_tx`
`serial_rx`

Exercise Solutions:
`Source_Code\Embedded\Exercises\CH10\`
1. `which_char`

Chapter 11

Example Programs:
`Source_Code\Embedded\CH11\`
`formats`
`escape`
`brk_cont`
`do_while`
`for_loop`
`nest_loop`
`nest_if`
`menu`
`sw_condition`
`point_func`
`return2`
`glob_var`
`static_var`

Exercise Solutions:
`Source_Code\Embedded\Exercises\CH11\`
1. `count_down`
2. `LED_select`

Chapter 12

Example Programs:
`Source_Code\Embedded\CH12\`
`index_array`
`LED_seq`
`strings`
`write_string`
`init_string`

`str_func`
`array_addr`
`pass_array`
`string_ptr`
`multi_seq`
`status_array`

Exercise Solutions:
`Source_Code\Embedded\Exercises\CH12\`
1. `sw_message`

Chapter 13

Example Programs:
`Source_Code\Embedded\CH13\`
`port_write`
`bitwise_port`
`not_port`
`knight_disp`
`or_keys`
`and_button`

Exercise Solutions:
`Source_Code\Embedded\Exercises\CH13\`
1. `num2bin`

Chapter 14

Example Programs:
`Source_Code\Embedded\CH14\`
`timer`
`analogue`
`timer_interrupt` (created from the **`Flash_interrupt`** template files)
`watchdog` (created from the **`Flash_template`** template files)

Chapter 15

Example Programs:
`Source_Code\Embedded\CH15`
`sample_struct`
`init_struct`
`point_struct`
`union`
`multibyte`
`enum_seq`
`Voltmeter`

```
Voltmeter_filt
stopwatch
```

Software Programs

The DJGPP compiler set up for PC programming with batch files:

The **ExtrasCprogramming.zip** file downloaded from www.elektor.com contains the **DJGPP.zip** file.

Copy the **DJGPP** directory in this zipped file to **C:**

If you would like to send comments or suggestions to the author about this book, please e-mail me at:
elektor.c.programmming@gmail.com

Index